Equality Is a Struggle

Equality Is a Struggle

Bulletins from the Front Line, 2021–2025

THOMAS PIKETTY

Yale UNIVERSITY PRESS
New Haven and London

English translation © Thomas Piketty and *Le Monde*, 2025. Originally published as *Vers le socialisme écologique: Chroniques 2020–2024*, © Éditions du Seuil, 2024.

All rights reserved.

This book may not be reproduced, in whole or in part, including illustrations, in any form (beyond that copying permitted by Sections 107 and 108 of the U.S. Copyright Law and except by reviewers for the public press), without written permission from the publishers.

Introduction translated by TranslationServices.com for Yale University Press. Chronicles from February 2021 to December 2023 translated by Kristin Couper for *Le Monde*. Chronicles from January 2024 to January 2025 translated by *Le Monde* in English.

Figures 1, 2, 3, 4, 11, 12, and 13 originally appeared in *A Brief History of Equality* by Thomas Piketty, translated by Steven Rendall, Cambridge, Mass.: The Belknap Press of Harvard University Press, Copyright © 2022 by the President and Fellows of Harvard College. Used by permission. All rights reserved.

Figures 5, 6, 7, 8, and 15 originally appeared in *A History of Political Conflict: Elections and Social Inequalities in France, 1789–2022* by Julia Cagé, Thomas Piketty, translated by Steven Rendall, Cambridge, Mass.: The Belknap Press of Harvard University Press, Copyright © 2025 by the President and Fellows of Harvard College. Used by permission. All rights reserved.

Figures 9 and 10 originally appeared in *Capital and Ideology* by Thomas Piketty, translated by Arthur Goldhammer, Cambridge, Mass.: The Belknap Press of Harvard University Press, Copyright © 2020 by the President and Fellows of Harvard College. Used by permission. All rights reserved.

Yale University Press books may be purchased in quantity for educational, business, or promotional use. For information, please email sales.press@yale.edu (U.S. office) or sales@yaleup.co.uk (U.K. office).

Set in Minion type by Westchester Publishing Services.
Printed in the United States of America.

Library of Congress Control Number: 2025938552
ISBN 978-0-300-28275-7 (hardcover)

A catalogue record for this book is available from the British Library.

Authorized Representative in the EU: Easy Access System Europe, Mustamäe tee 50, 10621 Tallinn, Estonia, gpsr.requests@easproject.com

10 9 8 7 6 5 4 3 2 1

Contents

Toward Ecological Socialism 1

Time for Social Justice 49

Combatting Discrimination, Measuring Racism 54

Rights for Poor Countries 58

From Basic Income to Inheritance for All 62

The G7 Legalizes the Right to Defraud 66

Responding to the Challenge of China with Democratic Socialism 70

Emerging from September 11 75

"Pandora Papers": Maybe It Is Time to Take Action? 79

Can the French Presidential Election Be Saved? 83

The New Global Inequalities 87

Rightward Shift, Macron's Fault 91

Sanction the Oligarchs, Not the People 95

Confronting War, Rethinking Sanctions 99

The Difficult Return of the Left-Right Divide 103

The Return of the Popular Front 107
Moving Away from Three-Tier Democracy 111
For an Autonomous and Alterglobalist Europe 115
A Queen with No Lord? 119
Rethinking Federalism 123
Redistributing Wealth to Save the Planet 127
Rethinking Protectionism 130
President of the Rich, Season 2 134
Emerging from the Pension Crisis through Justice and Universality 138
Macron, the Social and Economic Mess 142
Can We Trust Constitutional Judges? 146
What If Economists Were about to Change? 150
For a European Parliamentary Union (EPU) 154
France and Its Territorial Divides 158
Who Has the Most Popular Vote or the Most Bourgeois Vote? 162
Israel-Palestine: Breaking the Deadlock 167
Taking the BRICS Seriously 171
Escaping Anti-Poor Ideology, Protecting Public Service 175
Rethinking Europe after Delors 179
Peasants, the Most Unequal of Professions 183
When the German Left Was Expropriating Princes 187
Should Ukraine Join the EU? 191

For a Binational Israeli-Palestinian State 195
For a Geopolitical Europe, neither Naive nor Militaristic 199
Rebuilding the Left 203
Europe Must Invest: Draghi Is Right 207
How to Tax Billionaires 211
Unite France and Germany to Save Europe 215
For a New Left-Right Cleavage 219
Democracy vs. Oligarchy, the Fight of the Century 223

Index 227

Toward Ecological Socialism

The twentieth century was the century of social democracy. The twenty-first century will be that of ecological, democratic, and participatory socialism. This statement may come as a surprise at a time when a worrying mix of identity-based withdrawal and resigned neoliberalism seems to be gaining ground everywhere. And yet I remain optimistic. As I tried to show in *A Brief History of Equality*, equality is a struggle, and above all, it's a struggle that can be won, that was won in the past, and will be won in the future.[1] But that can only happen on the condition that we take full measure of the institutional transformations that it implies, learn all the lessons from the political strategies that come out of it, and never leave social and economic questions and reflections on the alternative socio-economic system to others. These are the eminently political questions that all citizens should have an opinion on and get involved with. It's by overturning the relations of knowledge and power, and by resuming the course of social and collective mobilization, that

---

[1] T. Piketty, *A Brief History of Equality* (Harvard University Press, 2022).

the march toward equality and dignity will be resumed and the national-liberal parenthesis will once again be closed.

No Habitable Planet without Egalitarian Decommodification

First things first: None of the social, environmental, and global challenges that we face today will be resolved without a drastic reduction of inequalities at a global level and a deep questioning of today's market and capitalist logics. In other words, democratic and ecological socialism will end up imposing itself because the other ways of thinking—starting with liberalism and nationalism—will never be able to resolve the challenges of our time by themselves. Electoral democracy needs a strong socialist and egalitarian pillar to function, and it's this pillar that has been lacking since the 1980s and 1990s, which in large part explains the current political dysfunction, as well as our collective inability to respond to global challenges.

To avoid climate collapse, production and consumption patterns will have to be profoundly changed for all social groups and regions of the world. But the working and middle classes of the Global North and the Global South alike will never accept the necessary changes if we don't start by demanding even greater effort from the most privileged social classes—and particularly from billionaires and other multimillionaires who like to lecture others while their carbon emissions and their threats to the planet's habitability are considerably larger than those of the rest of the population. As the number of environmental catastrophes increases, this reality will be increasingly evident and will end up radically changing attitudes toward the current capitalist system and the vast inequalities that it generates.

We can turn things on their head, but the facts remain: However we measure these realities, the world's richest 10% are responsible for a disproportionate share of global carbon emissions, incomparably greater than the poorest 50% or the following 40%, as the World Inequality Lab has shown. If we break down emissions according to investment and capital ownership, meaning if we consider that, above all, it is the owners of financial and real estate capital who are responsible for the technological and productive choices made and the resulting emissions, then the richest 10% are responsible for around 70% of emissions, compared to barely 5% for the poorest 50%. Conversely, if we break down emissions according to consumption by different social groups, an idea that could be justified but that undoubtedly tends to overestimate citizen-consumers' ability to influence the ecological content of the goods and services offered to them, then the richest 10% are responsible for around 40% of emissions, compared to 20% for the poorest 50%. And if we take the intermediary point of view by weighting consumption and investment according to their share of national income and emissions, the richest 10% account for nearly 50% of emissions, compared to 15% for the poorest 50%.[2]

Whichever way you look at it, the fact is that the concentration of emissions is extremely high, and the same is true for environmental degradation as a whole. The richest have a disproportionate responsibility for climate change: This

---

2 See the *World Inequality Report 2022*, World Inequality Lab/Seuil, wir2022.wid.world. See also L. Chance, P. Booth, and T. Voituriez, *Climate Inequality Report 2023*, inequalitylab.world; L. Chancel and Y. Rehm, "The Carbon Footprint of Capital," World Inequality Lab, Working Paper 2023/26, https://wid.world/news-article/the-carbon-footprint-of-capital/, December 2023.

particularly applies to the richest in the West, China, Russia, India, the Middle East, and other regions. This is why solutions lie in a global reduction in inequalities between social classes, not simplistic and reductive opposition between nation-states (which are by no means homogeneous within themselves). A massive reduction in carbon emissions and other ecological harms by the richest is a sine qua non for limiting climate change and keeping the planet habitable, both because of the wealthiest's considerable share in total emissions and harms and because it's impossible to involve other social groups in the transformation of lifestyles and production patterns if this minimum condition of justice and coherence is not met.

However, although reducing inequalities is a necessary condition for ensuring the planet's habitability, it is in no way sufficient. The drastic compression of wealth disparity—let's set a maximum scale from 1 to 5 for income and from 1 to 10 for wealth—would certainly be an excellent thing in itself, regardless of the ecological issue. All the historical and comparative data collected by the World Inequality Database (wid.world) suggest that such a compression is possible and collectively desired. But it would be erroneous to think that reducing inequalities is enough to guarantee sustainable development and a habitable planet by 2100. A world that is perfectly equal but where everyone continues to be as dependent on hydrocarbons, plastic, and concrete as they are today would also not be particularly desirable. What we need today is, above all, a process of egalitarian decommodification, meaning a drastic reduction in inequalities that would also allow for a gradual and resolute exit from market and capitalist logic in an ever-growing number of business sectors, and eventually the entire economy. In concrete terms, this signifies that entire sectors, starting with energy, transportation, and construction, must move away from a logic of pure profit. This can be achieved through a multitude of players, ownership regimes, and participatory and democratic governance, but it always involves respecting extremely strict common public standards (banning gas engines and plastic for most uses, construction and insulation standards, etc.), with dissuasive penalties for those who do not comply.

Decommodification Has Started: The March toward Equality in the Twentieth Century

The good news is that this process of egalitarian decommodification has already largely started and was even an immense success in the twentieth century. In large part, the construction of the social state and the triumph of social democracy in the twentieth century can be analyzed as a particularly successful process of egalitarian decommodification. To take the measure of the institutional transformations involved, let's start by recalling that all compulsory levies (all levies together, including direct and indirect taxes, social security contributions, etc.) were less than 10% of national income in Europe on the eve of World War I and have risen to 40–50% of national income since the 1980s and 1990s. In the nineteenth century and until 1914, the state was content with ensuring traditional sovereign functions (order and security). Social expenses, particularly education and health, were almost entirely absent. Then, throughout the twentieth century, first during the interwar period and especially in the decades following World War II, public authorities gradually took on a diversified and increasingly wide range of public service and social

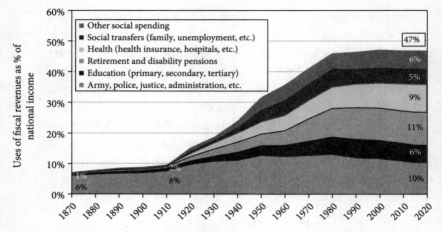

Figure 1. The rise of the social state in Europe, 1870–2020. In 2020, fiscal revenues represented 47% of national income on average in Western Europe and were used as follows: 10% of national income for regalian expenditures (army, police, justice, general administration, basic infrastructure, roads, etc.); 6% for education; 11% for pensions; 9% for health; 5% for social transfers (other than pensions); 6% for other social spending (housing, etc.). Before 1914, regalian expenditures absorbed almost all fiscal revenues.

Note: The evolution depicted here is the average of Germany, France, Britain, and Sweden.

Sources and series: piketty.pse.ens.fr/equality, figure 19.

protection missions: schools, hospitals, housing, transportation, and pensions (see figure 1).[3]

This unprecedented transformation, which we can describe as a "social-democratic revolution," is a significant

---

[3] The evolution shown in figure 1 corresponds to the mean evolution observed for Germany, France, the United Kingdom, and Sweden. There are significant differences in levels and chronologies between European countries, but concerning long-term trends, the various national trajectories appear relatively similar. See T. Piketty, *Une brève histoire de l'égalité* (Seuil, 2021), pp. 178–187, and *Capital et idéologie* (Seuil, 2019), pp. 534–538.

historic event. If the liberals and conservatives of 1910 had been told that half of the national income would be socialized in the new century, they would have undoubtedly warned of the red peril, the collectivist hydra, the economic collapse that would result. Not only did this collapse not happen, but the twentieth century was in fact characterized by unprecedented economic prosperity, driven by a previously unheard-of work productivity (wealth produced per hour worked) and by a very significant compression in income gaps, and to a lesser degree those of wealth.[4] There was one simple reason for this: The key to prosperity lies above all in the widest and most inclusive possible access to human capital (in particular to education and healthcare) and to collective infrastructures, and certainly not in the hyperconcentration of ownership and class privileges that characterized European societies before 1914.

The main point I wish to insist on here is that the construction of the social state in the twentieth century is inseparable from a process of decommodification of large sectors of the economy. In concrete terms, entire sectors such as education, healthcare, research, social security, and to a lesser degree transportation, energy, and housing developed outside the classic lucrative capitalist logic in most countries in the twentieth century, particularly in Western and Northern Europe. Goods and services of considerable importance were produced by a diversified group of non-capitalist players (public administrations, local authorities, associations, municipal companies, social security funds, schools and universities, hospitals and care facilities, regulated and contracted professions), eventually with excellent results in terms of collective efficiency and public health indicators. By contrast, in the United States,

---

4 See Piketty, *Une brève histoire de l'égalité*, figures 6–7, pp. 65–74.

where the social state has grown less rapidly than in Europe (notably because of the weight of racial antagonism), and where the healthcare system in particular relies very heavily on the lucrative private sector, we can see that the total cost of the healthcare system (public and private) represents a much larger share of national income than in Europe (close to 20% of national income in the United States, compared with 10–15% in Europe), for results that are clearly inferior according to all available indicators, and with the added bonus of abysmal inequalities.

However, the process of decommodification (to use Karl Polanyi's term) observed in the twentieth century remains incomplete, even in Europe.[5] About one quarter of the economy became non-market over the course of the twentieth century, but the other three quarters remained dominated by market, capitalist, and extractivist logics.[6] In all Western countries, as in the Soviet Union, Japan, and China, economic prosperity in the twentieth century and the early twenty-first century was notably constructed on an uninhibited exploitation of natural resources on a global scale, and in particular on a strategy

---

5 In his classic work *The Great Transformation,* published in 1944, the Austrian economist and historian Karl Polanyi shows how the process of "commodification" of the economy and the sacralization of the market and of competition in the nineteenth century up until 1914 contributed to weakening European societies and led to the disasters and destruction that followed. See Piketty, *Capital et idéologie,* pp. 490–492 and 548–555.

6 In absolute terms, with compulsory taxes approaching 50% of national income, it is possible to finance non-market sectors representing 50% of the economy; in practice, about half this tax income serves to finance monetary transfers (pensions, unemployment, and family allowances), not to produce goods and services, so non-market sectors (education, health, other public services, and public goods) represent about a quarter of the economy.

aimed at the unrestrained burning of hydrocarbon reserves accumulated in the earth over millions of years, with the well-known consequences of global warming and widespread environmental damage. Everywhere, the logic of short-term profit has prevailed over long-term consideration of general collective interest. Over the last few decades, the awareness of climate and environmental damage has developed slightly more strongly in Europe than elsewhere in the world, but the reduction of emissions nevertheless remains limited in comparison to historic responsibilities (especially if we include imported emissions), and initiatives to redistribute wealth globally remain in their infancy. Historically, European social democrats have sometimes emphasized reducing working hours (to the detriment of producing more, which should be welcomed), but their involvement in challenging consumerism and extractivism in general has remained rather inadequate.

Pursuing Egalitarian Decommodification in the Twenty-First Century

To summarize, the social-democratic revolution that took place in Europe in the twentieth century showed that it was possible to overcome capitalism and market and lucrative logics in a large number of business sectors, but unfortunately it has stopped in its tracks. We must reflect and build on this fundamental historical experience, with its successes and limits, to head toward a more ambitious and complete decommodification in the twenty-first century.

Three essential lessons emerge from historical analysis. The first concerns sectors where decommodification must be a priority in the twenty-first century, and the forms of democratic governance that must be developed. The second concerns the key role played by tax progressivity in opening up a new cycle of increasing the socialization of wealth. The third concerns the political strategy and social coalitions that will allow this social-democratic revolution to be pursued in the twenty-first century.

Let's start with the first point: decommodification sectors and their governance. The main challenge for the decades to come is that we must continue the growth of the non-market sectors of the twentieth century (notably education and healthcare, whose importance will continue to grow very strongly in the twenty-first century, regardless of whether we dedicate the necessary public resources to them or abandon the field to private and lucrative logics), all while giving ourselves the means to develop new ones (energy, transportation, construction and renovation, organic farming, and environmental protection in all its forms). The only way to face up to this double challenge is to rely resolutely on the increasing socialization of wealth, all while developing innovative modes of participatory, decentralized, and democratic governance in these numerous sectors. Nothing here can be considered as a given: The mobilization of new fiscal resources is still a fragile political process that taxpaying citizens can choose not to support at any moment; and the development of new organizational forms demands perseverance and humility and is in practical reality much more complicated than in a priori theories, especially since the structures must be constantly renewed and rethought in keeping with changing needs and demands for participation.

Taking the example of education, public resources have increased tenfold as a share of national income (less than 0.5% of national income before 1914 in Europe, compared with 5–6% of national income since the 1980s and 1990s), which has allowed the movement from a hyperelitist education system (where the vast majority of the population remains limited to a rudimentary primary education) to an unprecedented academic democratization, with almost all of the population reaching secondary education and now more than half of an age group reaching higher education. This impressive progression was achieved thanks to an institutional and organizational combination involving a large number of players: central public administrations, local authorities, schools and universities, middle schools and high schools, organizations representing teachers and parents, and others. For all that, these successes can never be taken as a given and must constantly be rethought and questioned in line with the ongoing democratization of education. For example, the diversity of needs and branches of higher education can justify a stronger decentralization and a greater autonomy for establishments (taking, for example, the forms of associations and foundations) than in primary and secondary education, all while reducing the inequality of access on social and local levels.

Similar problems are found in healthcare. Public resources in healthcare have grown even more rapidly (from less than 0.5% of national income before 1914 to about 10–12% today), allowing a speculative improvement of public healthcare indicators, with myriad players involved: central administrations, local authorities, social security funds, hospitals and clinics, contracted private practitioners, and so on. However, reflections on the ideal governance and organization of the sector are far from complete: an overhaul of hospital pricing and the doctor remuneration system, a greater role for nursing homes, greater involvement of patients and caregivers, and other factors.

The same questions are already being asked and will be asked more and more for the organization of new non-market sectors. The challenges are many: organization of local public transportation; production and distribution of renewable energy (wind, solar, hydro, biomass) at the local, national, and international levels; the management of water, forests, and natural resources; construction and renovation of buildings; promotion of local agricultural production and new forms of "social food security";[7] biodiversity protection; and so on. In all these sectors, innovative combinations of players will be required, including public authorities, groups of towns, municipal authorities, associations, and cooperatives. The solutions mostly need to be invented, based on the realization that lucrative and traditional capitalist logics cannot meet the challenges and needs and other organizational logics need to be found, patiently but with determination.

The process of decommodification and overcoming capitalist logic must also concern traditional lucrative market sectors. Throughout the twentieth century, social and trade union rights and the development of a salary status have, to some extent, balanced the powers between capital and work. Today, a business owner no longer has the same power to fire an employee or unilaterally change their salary as in 1910, just as a landlord no longer has the same power to replace a tenant or change the rent, and so much the better. In certain countries like Germany and Sweden, elected employee representatives have held a significant share of seats (between one third and one half) in the governing bodies of large enterprises (boards of directors or supervisory boards)

---

[7] See, for example, L. Petersell and K. Creteneis, *Régime général: Pour une sécurité sociale de l'alimentation* (Riot Editions, 2022).

since the 1950s. In concrete terms, this means that if employees hold an additional 20% or 30% of the company's shares, or if a local authority has such a stake, then it becomes possible to take control of the business and to outvote the shareholders who hold 70% or 80% of the shares. This represents a considerable change from the classic capitalist logic—one that could be imposed on shareholders only following intense social and political struggle. Finally, everything indicates that this system has allowed greater employee involvement in long-term investment strategies and has in no way harmed the economic prosperity of the countries in question—quite the opposite.

In absolute terms, there is no reason why such a system could not be more widespread, first by imposing it everywhere (and not just in Germanic or Northern Europe), then by extending it to small- and medium-sized enterprises (with the number of employee seats growing regularly with the size of the company), and finally, by limiting the number of votes that a single shareholder can hold in a large enterprise—for example, no more than 10% of the votes in a company with 100 employees (see figure 2). Recent debates have also seen the resurgence of discussions concerning the propositions of "wage-earner funds" created by Rudolf Meidner and his colleagues in the Swiss trade union federation LO in the 1970s and 1980s. According to this system, which concerns mainly the biggest companies, employers would be required to pay a portion of every year's profits into a wage-earner fund, allowing employees to gradually take control of 52% of the capital after twenty years.[8] Intended to

---

8 We could also imagine that the progressive wealth tax could be partially paid in the form of shares paid into wage-earner funds.

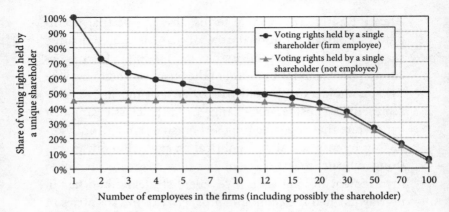

Figure 2. Participatory socialism and power sharing. In the system of participatory socialism, a single shareholder holding 100% of the firm's capital stock holds 73% of voting rights if the firm has two employees (including himself), 51% if the firm has ten employees (including himself), and he loses the majority if the firm has more than ten employees (including himself). A single shareholder who is not a firm employee holds 45% of the voting rights if the firm has fewer than ten employees; this share then declines linearly and reaches 5% with 100 employees.

Note: The parameters used here are the following: (1) employees (whether or not they are also shareholders) hold 50% of voting rights; (2) within the 50% of voting rights going to shareholders, no single shareholder can hold more than 90% of them (i.e., 45% of voting rights) in a firm with fewer than ten employees; this fraction declines linearly to 10% (i.e., 5% of voting rights) in firms with more than ninety employees (shareholder voting rights that are not allocated are reallocated to employees).

Sources and series: piketty.pse.ens.fr/equality, figure 18.

complement the existing system of employee voting rights (independent of any shareholding), this proposal was fiercely opposed by Swedish capitalists and could not be adopted. It was recently put back on the agenda by a number of US Democrats (in particular Bernie Sanders and Alexandria Ocasio-Cortez) and British Labour politicians.⁹ Other innovative proposals have also been formulated to allow the development of public investment funds at the local and communal level.¹⁰ The objective here is not to end the discussion but rather to show its full scope: Concrete forms of power, self-management, and economic democracy still need to be reinvented.

Opening a New Cycle of Increasing Socialization of Wealth

Let's recap. The first lesson of the twentieth-century social-democratic revolution is that the sectoral scope of decommodification cannot be predetermined any more than the multiple modes of governance that will have to be experimented with in order to move away from market and capitalist logic in an increasing number of business sectors. What is certain is that a new extension of non-market sectors will require the opening of a new cycle of increasing socialization of wealth, a prospect that must be clearly assumed. In the long term, should compulsory levies represent 60–70% or 80–90% of national income, and at what rate is such a trajectory of increasing socialization of wealth likely to occur? It is impossible to give a perfectly precise answer to this type of question in advance. In 1910, nobody could have foreseen that compulsory levies would rise from

---

9 See R. Meidner, *Employee Investment Funds: An Approach to Collective Capital Formation* (Allen & Unwin, 1978); G. Olsen, *The Struggle for Economic Democracy in Sweden* (Ashgate, 1992); J. Guinan, "Socialising Capital: Looking Back on the Meidner Plan," *International Journal of Public Policy* 15, no. 1/2 (2019): 38–58.

10 See J. Guinana and M. O'Neill, *The Case for Community Wealth Building* (Polity Press, 2020).

10% to 50% of national income in the new century. In the twenty-first century, as in the twentieth, everything will depend on the capacity of the public sector (in a general sense) and non-market logics to respond to citizens' concrete needs (education, healthcare, transportation, energy, housing, food, etc.) in a more convincing manner than the private sector and lucrative logics. It should also be emphasized that numerous public policies (monetary creation system, corporate governance rules, minimum and maximum salaries, bans on gas engines and plastic, construction and renovation standards, etc.) can have just as big an impact as taxes and transfers without affecting the share of compulsory levies in national income. The structural transformation of the socio-economic system is a multidimensional process that cannot be summed up with a single indicator. Finally, several different institutional combinations can achieve the same objectives, and only successful historical experiments can help in making decisions and progress. It's best to start from concrete needs and the organizational and financial questions that arise, sector by sector, while keeping in mind the multiplicity of long-term trajectories to which this process of overcoming capitalism can lead.

The second essential history lesson is that the increasing socialization of wealth in the twentieth century would not have been possible without the development of a very progressive tax system, meaning without the implementation of tax systems that apply much higher rates to those with very large incomes and wealth than to the rest of the population. This crucial role for tax progressivity by no means signifies that it is the richest who have significantly financed the social state. In practice, it's the entire population that was called to contribute throughout the twentieth century, and notably the middle and working classes, which, in all countries concerned, financed the social state mostly through direct taxes and social security contributions deducted directly from their salaries.[11] It's all the more obvious given that the period of accelerated development of the social state and compulsory levies from 1910 to 1990 was also marked by a major reduction in the wealth gap, and in particular by a fall in the share of very high incomes and wealth: So, it was not the top end of the distribution scale that had to be relied upon to finance the massive social effort.

While tax progressivity has nonetheless played a central role, this is due to a combination of factors. First, detailed analysis of available historical sources shows that the 80–90% rate imposed on very high incomes, such as in the United States between 1930 and 1980 (see figure 3), notably put an end to astronomical executive compensation packages (or at least made them much less widespread and much lower during this period), which contributed to freeing up significant room for growth for middle and lower salaries. These near-confiscatory rates that were applied to very high incomes did not undermine US growth or the country's prosperity, which is historical proof, if there ever was any, that very large salaries are of

---

11 In France, social security contributions and other social taxes like Contribution Sociale Generalisee (CSG) have historically played a much larger role than income tax (which until recently was not taken at the source). In other countries such as Denmark, social security contributions have actually played a negligible role: The majority of the social state was financed by a major income tax, the proceeds of which were destined for various social expenditures (pensions, healthcare, family, etc.). In every case, the income depends on the entire population, and in particular on workers in the lower and middle classes, not just on the upper classes.

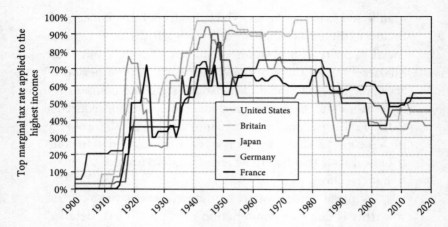

Figure 3. The invention of progressive taxation: the top income tax rate, 1900–2020. The marginal income tax rate applied to the highest incomes in the United States was on average 23% from 1900 to 1932, 81% from 1932 to 1980, and 39% from 1980 to 2018. Over these same periods, the top rate was equal to 30%, 89%, and 46% in Britain; 26%, 68%, and 53% in Japan; 18%, 58%, and 50% in Germany; and 23%, 60%, and 57% in France. Progressive taxation peaked in midcentury, especially in the United States and Britain.

Sources and series: piketty.pse.ens.fr/equality, figure 20.

no use to the general interest, and that prosperity depends predominantly on the level of training given to the workforce and the involvement of the largest number of people in the productive process (see figure 4).[12]

---

12 However, the fact is that over the period from 1930 to 1980, the United States had a very large educational advantage over Europe and Japan, resulting in a very large advantage in terms of work productivity, which the tax progressivity did not alter. See Piketty, *Une brève histoire de l'égalité*. For more detailed historical analysis of the role of tax progressivity,

Toward Ecological Socialism

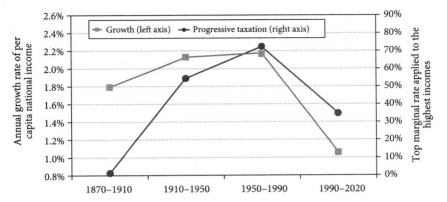

Figure 4. Growth and progressive taxation in the United States, 1870–2020. In the United States, the growth rate of per capita national income dropped from 2.2% per year between 1950 and 1990 to 1.1% between 1990 and 2020, while the top marginal tax rate applied to the highest incomes dropped from 72% to 35% over the same period. The promised resurgence of growth following the cut in top tax rates did not occur.

Sources and series: piketty.pse.ens.fr/equality, figure 23.

Then, in Europe and the United States, progressivity played an essential role in legitimizing the tax system and developing tax compliance, and therefore in the construction of the social state. In other words, during a large part of the twentieth century, the middle and working classes could be certain that the wealthy were contributing a significantly higher rate than themselves, making the effort demanded of them acceptable and legitimizing the increasing socialization

---

see T. Piketty, *Les hauts revenus en France au XXe siècle* (Grasset, 2001); *Le capital au XXe siècle* (Éditions du Seuil, 2013); and *Capital et idéologie*.

of wealth in its entirety. This is no longer the case, and this undermining of the progressive tax system since the 1980s and 1990s has weakened the entire social contract. If the richest pay lower taxes than I do, then why should I continue to pay for those less well off than I am? Step by step, the entire logic of social solidarity has been undermined. That is why it now seems impossible to begin a new phase of increasing socialization of wealth if we do not start by reestablishing an effective and incontestable tax progressivity that demands a real and substantial contribution from billionaires and other multimillionaires, as well as the most prosperous multinationals. To achieve this, it will be necessary to reverse the senseless system of free circulation of capital without tax relief put in place in the 1980s and 1990s, without the political players involved always understanding the consequences. In particular, the conditions will need to be created for collective ownership and social and trade union mobilization concerning a tax progressivity that is much stronger than that of the past, because it was this absence of ownership and mobilization that allowed the radical challenges of the 1980s and 1990s.

Finally, we cannot ignore another directly political effect of reducing inequality made possible by very strong tax progressivity: With the reduction of very high salaries and wealth in the twentieth century, the privileged classes' ability to influence opinions—notably through the financing of media and political campaigns—was weakened in the long term. Conversely, the increase in very high income and wealth at the end of the twentieth century and at the beginning of the twenty-first century has allowed a certain number of wealthy people to influence the political line of newspapers, television channels, think tanks, and other channels for disseminating dominant opinions. In a certain way, we find ourselves back in a situation already seen in the nineteenth century and at the beginning of the twentieth century, with a wealthy class bringing all its weight to bear on political processes and managing—at least for a time—to avoid redistribution for a long time after the disappearance of censal privileges. Today, as in the past, the solution is ambitious legislation aimed at democratizing the media by limiting the shareholders' power and developing egalitarian forms of financing political campaigns.[13] But to stop this legislation from being too frequently avoided, the indispensable democratic guarantee remains a very large reduction of income and wealth disparities, notably through a significant tax progressivity.

A Look Back at Revolutionary Social Democracy in the Twentieth Century

Let's summarize. The first lesson of historical analysis is that the range of non-market sectors cannot be predetermined. The second is that the increasing socialization of wealth demands a real and substantial tax progressivity and a large compression between income and wealth brackets. The third history lesson is perhaps the most important: It concerns the political strategy and social coalitions that made the twentieth-century social-democratic revolution possible and that can allow this trajectory to continue and increase in the twenty-first century. To sum it up, the social-democratic revolution was the fruit of a political strategy founded both on the conquest for power through electoral and parliamentary democracy and on social struggles as a spur to institutional transformation. The important point is that it would not have taken place without the development of powerful collective organizations (political

---

13 See J. Cagé, *Libres et egaux en voix* (Fayard, 2021).

parties and trade unions) to mobilize voters and workers around a programmatic platform aimed at overcoming capitalism and installing an alternative socio-economic system.

These political organizations and trade unions took different forms in the different countries involved, but they were all based on an agenda of radical transformation of the capitalist system and overcoming market and lucrative logics. In Sweden, the Social Democrats won the 1932 elections and held power almost continuously until the 1990s and 2000s. This allowed them to put in place a particularly developed social state and to completely transform the country, which, until the 1910s and 1920s, was one of the most inegalitarian and violently censal countries in Europe. In the United Kingdom, the Labour Party won an absolute majority for the first time in 1945, which allowed them to establish the National Health Service (NHS) and numerous social and tax policies that were previously unthinkable (e.g., tax rates above 95% on the highest incomes and wealth) in a country with a heavy aristocratic past and whose parliamentary system was dominated, until 1910, by the House of Lords (from which most prime ministers came). Both the British Labour Party and the Swedish Social Democrats are genuine workers' parties, intrinsically linked to trade unions, strikes, and factory jobs. In the eyes of the wealthy, it was the barbarians that were taking power! In 1944, the ultraliberal economist and philosopher Friedrich Hayek published *The Road to Serfdom*, in which he warns his Swedish and British readers: You must stop all compromise with the Social Democrats and the Labour Party, who will end up leading the country toward Bolshevik dictatorship with their totalitarian illusions of redistribution and social justice. The warning may make you smile today, especially as it came from someone who became one of the main intellectual supports of General Augusto Pinochet's military regime in the 1970s and 1980s.

But this also allows us to recall the fear that inspired social democracy in the middle of the twentieth century.

In France, the 1936 victory of the Front Populaire led to paid vacations and collective agreements. In 1945, the Communists and Socialists were in a strong position in the Assemblée Nationale and played a decisive role in establishing social security and major public services in the energy and transportation sectors. In Germany, the Social Democratic Party (SPD) played an essential role in developing the social state, not only during their time in power (in particular in the 1920s and 1970s) but also in opposition. Backed by a powerful trade union movement, they managed to put pressure on the Christian Democratic Union (CDU) to make them adopt the 1952 law allocating half the seats in major companies' governing bodies to employee representatives. Employers were up in arms. This was a general characteristic of the social-democratic revolution in the twentieth century: The various social-democratic, labor, socialist, and communist parties had managed to completely redefine the terms of the social and economic debate to the point that their political adversaries also ended up taking up their agenda. After the 1936 factory occupations, the Chambre des Députés unanimously adopted the Front Populaire's social measures, even though nobody was considering such decisions a few weeks prior. During the liberation, the Communists' and Socialists' influence was decisive, but the social program designed by the Conseil National de la Résistance also united a large part of the Christian Democrats and Gaullists. After 1968 and during the 1970s, in a frenzied social climate, it was the Gaullist and liberal governments that responded to trade union demands (a large increase in lower salaries, expansion of social security), before the Union de la Gauche in 1981, and throughout the following decades gave a new impetus to the French social state, notably concerning education (in particular with the objective of 80% of an age group obtaining the baccalaureate, which was launched in the 1980s and achieved three decades later) and healthcare (mainly with the universal health coverage created in 2000).

Finally, the clearest success of the social-democratic revolution is that it imposed its themes and political agendas: The idea of a social state funded by compulsory levies reaching 40–50% of national income, which would have seemed unthinkable at the start of the twentieth century, has become a clear goal for most political parties. Today, no political force in Europe is proposing to abolish public health insurance, free education, or social protections and revert to the situation of 1910, when levies represented less than 10% of national income. The issue in question is whether to freeze the weight of the social state at the level reached in the 1980s and 1990s (what right-wing and centrist parties are roughly proposing) or continue its extension and the historic process of socializing wealth (what a large proportion of left-wing parties are proposing, although not always in a coherent manner and without really succeeding to date). This state of the debate is far from satisfactory, but it illustrates to what point the construction of the social state in the twentieth century was a success that nobody wishes to undo (at least for the foreseeable future).

It should also be noted that the construction program for the social state, tax progressivity, and egalitarian decommodification supported in the twentieth century by the social-democratic parties (in a general sense, including the various nuances and coalitions of social-democratic, labor, socialist, and communist parties that came into power in Western and Northern Europe) did not correspond to a perfectly precise and predetermined agenda. At the start of the twentieth century, all these parties had a program aimed at a complete collectivization of the means of production. But the exact form of collectivization was not completely specified and could just as well include strictly state ownership and various forms of cooperative ownership and self-management. In practice, the essential difference with the Russian Bolsheviks (who initially were only the majority faction of the Russian Social Democratic Party, before being renamed the Communist Party in 1918) is that the Bolsheviks chose the "dictatorship of the proletariat" (a transition phase deemed inevitable by Karl Marx and Vladimir Lenin) and an authoritarian and repressive regime, whereas the European Social Democrats (including the various labor, socialist, and European communist parties) resolutely decided to inscribe their action within the framework of electoral and parliamentary democracy and social and trade union participation.[14] The compromises made within this framework took on the forms described previously, but it is clear that they could have taken on others. For the same reasons, it is impossible to predetermine the exact contours that egalitarian decommodification could take in the twenty-first century because, as in the past, it will rest on democratic participation, deliberation, and collective

---

14 I would go so far as to place the Parti Communiste Français among Europe's Social Democrats, a view that may indeed be contested but is not without justification. In November 1946, when the Communist leader Maurice Thorez was very close to becoming the head of government (the decisive vote was lost by a few dozen votes to dissident Socialists who refused to accept the party's decision to support the nomination of the Communist leader), he explained in a momentous interview with *Time* that his model of socialism had nothing to do with Soviet communism, first because it was firmly rooted in the French parliamentary tradition, and second because he formally promised to not touch small private property, particularly in the countryside. See J. Cagé and T. Piketty, *Une histoire de conflit politique: Elections et inégalités sociales en France 1789-2002* (Seuil, 2023), pp. 513–514.

experimentation. That is also why the social-democratic revolution proved to be more durable and revolutionary than the Bolshevik Revolution, which resulted in one of the worst kleptocratic regimes in history.

Rediscovering the Revolutionary Momentum of Social Democracy

Why has social democracy lost its subversive and revolutionary potential since the 1980s and 1990s, and under what conditions is it possible to begin a new phase of constructing the social state and the increasing socialization of wealth? To explain the social-democratic revolution's fading momentum, several factors have to be taken into account. We could evoke the exceptional circumstances that allowed the social state to develop in the twentieth century, in particular the two world wars, which exacerbated the social needs while also provoking an unusually strong weakening of the European ruling classes (much more so, for example, than in the United States, Latin America, or India), opening up the road to the European social-democratic revolution. This European revolution then lost its momentum once the initial crises had passed. However, the argument is too deterministic: The social-democratic revolution was first and foremost a global political response to numerous crises within industrial capitalism (colonial rivalries, the 1929 crisis, and more generally a social crisis in the making since the nineteenth century), and not simply a response to the wars. More generally, the social and financial crises return again and again in different forms (the 2008 crisis, Covid in 2020–2021, and especially the social and environmental crises of decades to come). The main question is how collective organizations can manage, or not, to transform these crises into majority social and political mobilizations and platforms for institutional transformation. It is first and foremost the political and organizational choices that explain the successes and failures of social democracy since the nineteenth century.

However, the fact is that, since the 1980s and 1990s, the various social-democratic parties (in a general sense) have stopped developing programs for the ambitious redistribution of wealth and overcoming capitalism. This can partly be explained by the success of the social-democratic program itself. Once the construction of an ambitious social state, based on compulsory levies representing 40–50% of national income, has been achieved (a level reached by the most advanced European countries in the 1980s and 1990s), it is tempting to consider the necessity to stop and concentrate on the consolidation and rationalization of existing social programs rather than on their indefinite expansion. The argument is enticing and has long seemed convincing to me, but in the end, it is insufficient and problematic. If we freeze the public resources available for education at 5–6% of national income, even though the share of university graduates from an age group has jumped from 20% to 50%, then we inevitably create a lot of frustration and inequalitiy, notably for all those who live far from large cities and large university centers. Similarly, if we freeze public healthcare resources at 10–12% of national income, despite the aging population and the constant development of new treatments (notably in the context of advanced technologies available in the hospital environment), then we will inevitably end up reducing the available resources for town doctors and ordinary care, especially for people living in small towns and rural areas. And if, generally, we froze the entirety of available public resources (as a share of national income) for all sectors together, while there are new primary needs, notably concerning the environment, housing, and energy and transportation infrastructures, then we inevitably create immense disappointments. An enormous gap between words and acts is slowly created, which is then very difficult to diminish. Here again, the sense of abandonment is particularly strong in small towns and rural areas, where public transportation is virtually nonexistent and where personal cars are widely used and very difficult to replace.

Social democracy is not a finished product: If we freeze ambitions at a given moment and explain that the only objective is to manage what already exists and defend the social triumphs of the past, without a real new perspective for the future, then we leave the field open to other political currents, particularly to supporters of the neoliberal freeze and the withdrawal into identitarianism. That is why, since the 1980s, 1990s, and 2000s, the main parties concerned—particularly the French Socialists, the British Labour Party, and the German Social Democrats—started to lose a growing portion of their working-class voters, notably in small towns and rural areas.

The Failure of Ecologism without Socialism

We must also emphasize the specific role of ecological movements and challenges in the weakening of the left since the 1980s and 1990s. I have noted here that future ecological catastrophes could help accelerate the criticism of capitalism and its overcoming. It is true, but only on the condition that the proposed overcoming rests on an ambitious project of democratic and ecological socialism (meaning on a platform based on the redistribution of wealth, the reduction of inequalities, and egalitarian decommodification), and not on an ecologism without socialism and aimed at the privileged classes.

The entire problem of ecological parties since their emergence in the 1970s and 1980s, particularly in France and Germany, is precisely that they have never put the social question at the heart of their message. Their defense of the environment is often formulated as if social classes did not exist, or, at least, without placing the class structure and the need to redistribute wealth and reduce social inequalities at the center of their analysis. The problem is that environmental measures designed without explicitly taking social class into account almost inevitably backfire on the working class in practice. A typical example is the carbon tax: If we raise the price of fuel and energy in the same proportion for all social classes, then in practice, this signifies that the effort required represents a much larger proportion of income for low- and middle-income classes than high-income classes because, on average, the former dedicate a much larger proportion of their income to these expenses. This is notably the case for the working classes in rural areas and small towns. If we add the fact that the richest benefit from numerous exemptions to this (e.g., jet fuel), and they can also evade taxes by billing their travel and consumption on company accounts, then the carbon tax appears as a caricatured example of a regressive tax, meaning that it weighs much more heavily on the poorest than on the richest, like consumption taxes and other indirect taxes in the nineteenth century, and in contrast to the progressive taxes on high incomes and wealth developed throughout the twentieth century, which have been undermined since the 1980s and 1990s. More generally, ecologism without social classes and without massive redistribution of wealth and economic power often seems like a trick on the working classes.

If we examine the voter profile for green candidates that have succeeded each other in French presidential elections

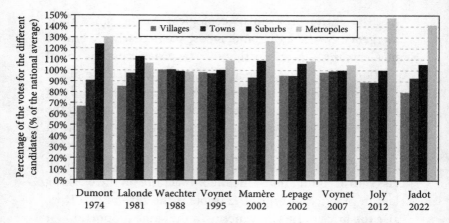

Figure 5. Political ecology and the territorial cleavage, 1974–2022. The representatives of political ecology in the presidential elections conducted from 1974 to 2022, whether it is Dumont (1% of the vote), Lalonde (4%), Waechter (4%), Voynet (3%), Mamère (5%), Lepage (2%), Voynet (2%), Joly (2%), or Jadot (5%), have almost always achieved higher scores in the metropoles and suburbs than in the towns and villages, with even an acceleration of this tendency toward the end of the period.

Sources and series: unehistoireduconflitpolitique.fr, figure 12.22.

between 1974 and 2022, we can note two striking patterns. On the one hand, the green vote is characterized by a particularly marked territorial divide, with much higher scores in large towns than in the smallest ones (see figure 5).[15] Moreover,

---

[15] The cities as well as the suburbs shown in figure 5 comprise the principal and secondary municipalities of conurbations with more than 100,000 inhabitants; the large villages comprise the municipalities of conurbations with between 2,000 and 100,000 inhabitants; and the villages comprise the municipalities of conurbations with less than 2,000 inhabitants. These four categories have the advantage of dividing the French population into four groups of comparable size for the recent decades.

Toward Ecological Socialism

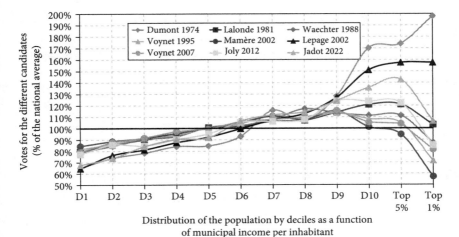

Figure 6. Political ecology and wealth, 1974–2022. In the 1974 presidential election, the vote for Dumont was a sharply increasing function of the municipality's income throughout the distribution. Subsequently, the vote for the ecological candidates was generally an increasing function of municipal income, except for the richest municipalities.

Note: The results indicated here are after controls for the size of the conurbation and municipality.

Sources and series: unehistoireduconflitpolitique.fr, figure 12.23.

the green vote systematically grows with the town's wealth, including for a given town size, except within the richest 5% of towns (see figure 6).[16]

---

16 The same general results can be seen with the individual data from post-electoral surveys, which, due to their small sample size, are less suitable for finely cross-referencing territorial and wealth divides than communal data. See Cagé and Piketty, *Une histoire du conflit politique*. See also A. Gethin, C. Martinez-Toledano, and T. Piketty, *Clivages politiques et inégalités sociales: Une étude de 50 démocraties (1948–2020)* (Seuil/Gallimard/EHESS, 2021) (wpid.world).

These two regularities (especially the second) are major anomalies compared to the usual historical pattern of voting for the various left-wing parties—namely, Socialists, Communists, Radicals, Trotskyists, or La France Insoumise (LFI). Generally, the vote for left-wing parties has definitely tended to be higher in urban environments than in rural environments for all elections held in France since the nineteenth century (with a few exceptions, like the Trotskyist and Radical votes). However, together with Julia Cagé, I have shown that this territorial divide between the left and the right was relatively smaller during most of the twentieth century.[17] De facto, the social divide outweighed the territorial divide between 1910 and 1990. The left-wing parties had managed to develop an ambitious redistribution platform and convince the urban and rural working classes that what brings them together is more important than what divides them, therefore opening the way to left-right polarization and to a "classist" political conflict centered on the reduction of social inequalities and the construction of the social state. By contrast, between 1990 and 2024, the territorial divide reached extremely high levels that had not been seen since the end of the nineteenth century and the start of the twentieth century. This allowed an objectively very socially privileged central liberal bloc to occupy power on the basis of the divisions of the urban and rural working classes between the left and the right. In this whole scheme, the emergence on the left of a particularly urban green vote significantly contributed to this transformation.

The second regularity is even more problematic, given that all left-wing votes have always had a clearly decreasing profile with the town's wealth. This has historically been the

---

17 See Cagé and Piketty, *Une histoire du conflit politique.*

case for Communist, Trotskyist, and LFI votes, but it has also been the case for Socialist and Radical votes, which until now have systematically had decreasing profiles with the town's wealth (if only slightly). It is particularly striking to note that the Trotskyist vote has always had a profile that is the exact opposite of the green vote in terms of both the territorial divide and the wealth divide (see figures 7–8). It is not about idealizing the platform of socio-economic transformation promoted by the Trotskyist organizations, which has clear limits and limited electoral success (albeit comparable to the success of the Greens in presidential elections over the past few decades), but simply

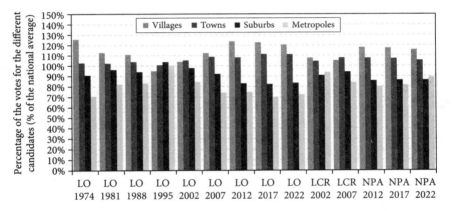

Figure 7. Trotskyism and the territorial cleavage, 1974–2022. The Trotskyist party LO was represented in the presidential elections by Arlette Laguiller in 1974 (2% of the vote), 1981 (2%), 1988 (2%), 1995 (5%), 2002 (96%), and 2007 (1%), and by Nathalie Artaud in 2012 (1%), 2017 (1%), and 2022 (1%). The LCR was represented by Olivier Besancenot in 2002 (5%) and 2007 (4%), and the NPA by Philippe Poutou in 2012 (1%), 2017 (1%), and 2022 (1%). The votes for LO, LCR, and NPA were systematically higher in the villages and towns than in the suburbs and metropoles for all these elections, except for the LO's initial breakthrough in 1995.

Sources and series: unehistoireduconflitpolitique.fr, figure 12.24.

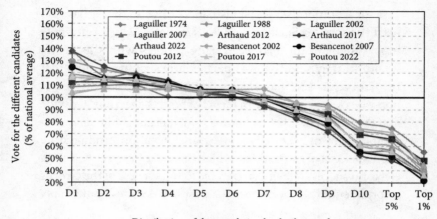

Figure 8. Trotskyism and wealth, 1974–2002. From the presidential elections of 1974 to those of 2022, the Trotskyist candidates have systematically had a vote profile that sharply decreases as a function of municipal wealth, whether it is a question of the LO candidates (Arlette Laguiller and Nathalie Arthaud), the LCR candidate (Olivier Besancenot), or the NPA candidate (Philippe Poutou).

Note: The results indicated here were calculated after controls for the size of the conurbation and municipality.

Sources and series: unehistoireduconflitpolitique.fr, figure 12.25.

about noting the fact that a greater or lesser emphasis on social inequalities and the redistribution of wealth has a massive impact on the characteristics of the voters who identify with the various discourses. In this case, over the last half century, the green discourse has had the greatest difficulty in attracting working-class voters, especially in rural areas, but also in urban areas. In practice, these voters often feel stigmatized for their environmental responsibilities (e.g., with their use of personal cars or detached houses), while the relatively favored urban classes that stigmatize them are often responsible for much greater environmental damage (e.g., through their use of planes or, more generally, their higher incomes).

From Social Democracy to Democratic and Ecological Socialism

To get away from these contradictions, the only solution is for all varieties of left-wing parties, allied with green parties, to develop an ambitious twenty-first-century program of wealth redistribution and egalitarian and ecological decommodification, along the lines of the social-democratic revolution of the twentieth century. To create this new horizon, it seems preferable to talk about "democratic and ecological socialism," but, of course, several terms are also conceivable. Some would opt for "eco-socialism." Others would prefer to talk about "social democracy for the twenty-first century." So long as we can agree on the content, the question of terminology can be considered secondary. In fact, every country, every language, every region of the world has its own history with these terms, which must not be fetishized. In Switzerland, the party designated as Social Democrats in German (Sozialdemokratische Partei der Schweiz) is called the Parti Socialiste Suisse (Swiss Socialist Party) in French.

Some researchers believe that "social democracy" is the ultimate horizon of our times, and that we must be wary of the notion of "democratic socialism" and the dangerous and useless illusions that these terms imply.[18] I understand this point of view, but the problem is that it most commonly rests

---

18 See, for example, L. Kaneworthy, *Would Democratic Socialism Be Better?* (Oxford University Press, 2022).

on a static and frozen vision of "social democracy," considered a near-finished product to defend, not a dynamic process in evolution and permanent renewal.[19] The same is true for public debates, notably in France, where numerous political actors close to the central liberal bloc use the term "social democrat" to refer to a relatively conservative agenda (which roughly consists of freezing public spending at its current level as a share of national income), in opposition to the so-called radical left and its vain promises.[20] Finally, this conservative and instrumental use of the term "social democrat" implies turning one's back on the revolutionary, subversive, and popular dimension of social democracy in the twentieth century. For me, the advantage of the notion of "democratic socialism" is that it clearly expresses the idea that it is about pursuing the social-democratic revolution of the twentieth century and assigning it new objectives concerning the struc-

---

19 In this instance, Kaneworthy defines "social democracy" as a socioeconomic system where the compulsory levies reach about 50% of national income and non-market employment about 25–30% of total employment (roughly equivalent to the levels already observed in Northern Europe). By comparison, he defines "democratic socialism" as a system where two thirds of employment and production would take place in organizations, administrations, and companies owned or controlled by the state, citizens, or workers. Kaneworthy considers such an objective as unrealizable (and undesirable) without specifying the ownership or control thresholds used, nor the boundary between an ordinary "citizen" and a major shareholder. Following the thresholds used, it is not impossible that "democratic socialism" has already been achieved in numerous countries, particularly in Germanic and Nordic Europe, if we take into account the votes of employee representatives, small shareholders, and public authorities.

20 In Brazil and Portugal, the Social Democratic Party positions itself as center-right and opposes the Workers' Party (Brazil) and the Socialist Party (Portugal), which are center-left. This once again illustrates the very large international variations in the usage of these terms, which should not be sacralized.

tural transformation of the socio-economic system, with the continuation of the egalitarian decommodification process and its gradual extension into new sectors, and eventually the whole of the economy at its core. While it is clear that such an agenda will never be achieved overnight, it nevertheless seems to me essential to start publicly debating the alternative socio-economic system that we want to put in place to respond to the global challenges of the twenty-first century and to move away from the frozen vision of social democracy that emerged in the 1980s and 1990s.

From Syriza to the Nouveau Front Populaire via Sanders: Hope and Restructuring on the Left

Let us conclude. The call for democratic and ecological socialism outlined here does not only respond to a long-term historical necessity. It also corresponds to a political evolution that has largely already started, and that everything indicates will continue. Between the fall of Soviet communism in 1990–1991 and the financial crisis of 2008, the world certainly knew a brief but intense period of liberal euphoria. Just about everywhere, people thought that all the world's problems could be resolved by liberalizing all markets on a global scale, by implementing the free circulation of goods and capital without any control, and by ignoring the social and environmental consequences of these decisions. The 2008 financial crisis sounded the death knell for these illusions: Without resolute action by the public authorities to absorb losses and stabilize the financial system, the banking sector would have collapsed and led the world to an economic collapse comparable to that of the 1930s. The 2020 health crisis, and the growing awareness of the seriousness of the environmental crisis, have also contributed to discrediting the unbridled liberalism of the 1990s and 2000s. Over the past fifteen years, numerous political movements have been seeking to propose new ways for overcoming capitalism.

The closest example to the democratic and ecological socialism defended here is undoubtedly embodied by the candidacies of Sanders in the 2016 Democratic primaries and then Sanders and Elizabeth Warren in the 2020 Democratic primaries. Despite notoriously unfavorable treatment in the media, these two candidates were virtually on par with Joe Biden, whom they clearly outpaced among voters under age 50. They developed a platform founded explicitly on economic democracy, the election of employee representatives to boards of directors (a revolution on the other side of the Atlantic), universal public health coverage, a marginal rate of more than 70% for the highest incomes in order to finance a major investment in public universities and cancel student debt, a progressive wealth tax up to 8–10% for billionaires (much more than all the wealth taxes applied in Europe in the twentieth century), all with a 40% exit tax on the assets of US taxpayers who choose to leave the country (which would be equivalent to a radical rethinking of the free circulation of capital). Although Biden won by a hair, on a platform that was clearly more centrist than that of Sanders and Warren (but still more interventionist than that of Barack Obama and Bill Clinton, the previous two Democratic presidents), the fact is that that election could have turned out completely differently, depending on more or less contingent campaign events. The Democratic Party's future elections could also turn out differently, with considerable consequences for the entire global political landscape. The line taken by Kamala Harris in 2024 seemed closer to Biden's, but nothing is set in stone for the future.

Almost at the same time as Sanders and Warren, Labour Party member Jeremy Corbyn was defending a line that turned its back on the Tony Blair–Clinton years and only just lost the 2017 British elections, with 40% of the vote compared to 42% for the Conservatives. During the 2024 elections, Labour member Keir Starmer won with a more centrist line than Corbyn's. But the fact is that he received less than 34% of the vote, with participation of the general electorate also falling drastically (68% in 2017, 59% in 2024), which shows that many voters (especially those from the working class) who had voted for Corbyn preferred to abstain. In practice, Starmer won only thanks to the division on the right between the Conservatives (with their liberal and pro-business line in disarray, getting only 24% of the vote) and the anti-migrant nationalists of the Reform Party (14% of the vote). This shows that although it is not easy for the proponents of democratic socialism to gather a large enough voter bloc, the task is just as difficult (if not more so) for the defenders of these two other main thought systems (liberalism and nationalism). In particular, the followers of a purely liberal and pro-business line (such as the Tories under Rishi Sunak and the Macronian bloc in France) invariably find themselves with a narrow and objectively very privileged voter base.[21]

Aside from the case of the US Democrats and the British Labour Party, since the 2008 crisis we have seen attempts just about everywhere within left-wing parties (Social Democrats, Labor, Socialists, Communists, etc.) to promote platforms of more ambitious socio-economic transformation than in the past. In continental Europe, these attempts have often been met with several structural difficulties, particularly

---

21 See "Who Has the Most Popular Vote or the Most Bourgeois Vote?," September 19, 2023. (References indicated with only a title and date are articles by Thomas Piketty that were published in lemonde.fr/blog/piketty and appear in this volume.)

linked to the historical fragmentation between small countries (engaged in a mutual competitive process and tax and social dumping that are difficult to deal with in isolation, both in Northern and Southern Europe) and within countries between rival political organizations. In Greece, the left-wing coalition Syriza won the 2015 elections on a more leftist line than the historic socialist party (Pasok) and permanently replaced it. Worried that the citizen coalition Podemos would do the same to the Spanish socialists of the Partido Socialista Obrero Españo (PSOE), which they came close to doing, and more generally worried about a bigger swing to the left, European leaders, particularly those of France and Germany, decided to impose drastic financial conditions for Greece and refuse major rescheduling of debt payments (despite previously promising to do so). The strategy certainly discredited the Syriza experiment and limited the shift to the left, but it also reinforced the nationalist right (which came into power ten years later in several European countries, notably in Italy).

Generally, attempts to reorganize the left undertaken in various European countries bear the mark of multiple national histories and are often weakened by divisions between political movements and by organizational and ideological bricolage. In Italy, the political landscape remains marked by the collapse of historical parties in 1991–1992. Among the new structures that emerged, M5S (the Five Star Movement) has a large working-class electorate, but it is characterized by an unstable territorial identity and a hesitant left-wing anchoring, as shown by its experience in power with the nationalist right before turning again toward the left. On the French political scene, LFI easily capitalizes on the repeated disappointments caused by the Socialists in power (twenty years in government between 1981 and 2017), all while applying a seemingly radical rhetoric and dubious democratic practices. In practice, the performance of the alliance of left-wing and green parties, the Nouvelle Union Populaire, Écologique et Social (NUPES) in 2022 and the Nouveau Front Populaire (NFP) in 2024, remains far too modest to obtain an absolute majority in the Assemblée Nationale, despite strong voter turnout and profound doubts aroused by the liberal Macronian bloc and the nationalist Rassemblement National bloc. These limited results can first and foremost be explained by the insufficient work done on the program and by the inability to establish a common democratic structure capable of organizing deliberations and settling disputes.[22]

In Germany, in the 2021 elections, the Left Party achieved a result that was far too low to hope to form a majority coalition with the social democrats of the SPD and the Greens, so the latter had no other choice than to call on the liberals of the Free Democratic Party (FDP) to complete their coalition (which numerous Social Democrat and Green members of the parliament would have probably preferred to do anyway). Since then, the landscape has become even more complicated: The three parties in power have seen their popularity decline; Alternative for Germany (AFD), on the anti-migrant and nationalist right, has grown in power; the CDU has regained its strength; and a new party that split from The Left—the Bündnis Sahra Wagenknecht–Vernunft und Gerechtigkeit (BSW; Sarah Wagenknecht Alliance)—has made a breakthrough in the 2024 European elections, as well as in several regional elections in the former East Germany. Anchored on the left by its history and its leaders' trajectories, the BSW movement predominantly presents itself as a defender of neglected territories, workers, and Germany's industrial network, all while

22 See "Rebuilding the Left," July 13, 2024.

concentrating a major part of its attacks against the well-to-do urbanites who vote for the Green Party (and to a lesser degree for the Social Democrats).[23] Its ambiguous remarks on immigration also contribute to complicating any prospect of a union of left-wing forces on the visible horizon.

To recap, the risks of a lasting fracturing on the left are serious all throughout continental Europe. The renewal of political organizations (some of which are of a venerable age) can be necessary and legitimate, provided that it is an opportunity to develop new democratic organizations that can unite large segments of the electorate around a real platform of transformation with a majority. The resentment against historic parties does not lead anywhere, but neither does wallowing in very low scores. However, it seems to me that this political excitement observed on the left throughout these fifteen years, in the United States, the United Kingdom, and most continental European countries, is promising. In its own way, it expresses a need to enter a new cycle and to extend the twentieth-century social-democratic revolution into the twenty-first century.

And If Political Innovation Came from India or Brazil?

It is certainly not my aim to claim that the road to democratic and ecological socialism is clearly marked: Everything, or almost everything, remains to be invented, particularly the urgent need to rethink an internationalism that finally turns its back on the sacralization of markets and allows for another way of sharing wealth on a global scale.[24] However, several

---

23 See S. Wagenknecht, "Condition of Germany," *New Left Review*, 2024.
24 See "Reconstructing Internationalism," July 14, 2020. See also "For an Autonomous and Alterglobalist Europe," July 12, 2022; "Rethinking

points need to be clarified. First, the march toward democratic and ecological socialism will necessarily take place over decades, just as the twentieth-century social-democratic revolution did, with moments of energizing acceleration and others of stagnation, or even regression. The construction of new collective organizations (political movements and trade unions) capable of carrying such a transformation is undoubtedly the most crucial challenge, and also the most complex and the slowest. Then, although the task is arduous for democratic and ecological socialism, other thought systems—starting with liberalism and nationalism—also face considerable contradictions, which, in my opinion, are even greater. Once in power, the liberal and nationalist forces very quickly lose a large part of their voter base for the simple reason that the responses they propose to the great challenges of our time are often insufficient, even counterproductive. Let us repeat: Democratic and ecological socialism will return time and time again because the other thought systems will never manage to resolve the challenges of our times on their own.

Finally, and perhaps most important, it would be deeply erroneous to expect political innovation to come solely from Europe and the United States, even if there are real possibilities for promising developments in both cases. It is entirely possible that the most decisive innovations in the coming decades will come from Africa, Asia, or Latin America, and more generally, that the movement toward democratic and

---

Federalism," October 11, 2022; "Rethinking Protectionism," December 13, 2022; and "For a European Parliamentary Union (EPU)," June 13, 2023, in this volume. This is undoubtedly the theme that comes up most frequently in my columns, not without repetition, and also some shifts in thought as events and public discussions on these issues evolve.

ecological socialism will come as much from the South as it does from the North. In India, a country that has more voters than all Western democracies combined, and that will perhaps become the world's leading economic and political power in the twenty-first century, an electoral cycle has undoubtedly started to close with the 2024 elections. The nationalist Hindus of the Bharatiya Janata Party (BJP) have registered a sharp decline, and it is entirely possible that the Congress Party allied with the socialist and communist parties to its left (which have a rich and original history in the context of India) will gain power at a federal level in the next election or the following one. Given that India is one of the countries of the world that will suffer and have already suffered the consequences of climate change most intensely, we can imagine that this will lead the country to take strong initiatives to redefine the rules of the global economic and financial system, such as concerning the governance of the International Monetary Fund (IMF), the World Bank, or the United Nations, or even implementing a minimum tax on multinationals and billionaires to finance climate repairs on a global scale. Very close to the business world, the BJP government has so far not played a major role in promoting such an agenda on an international level. It could be different with an Indian government led by the Congress Party or the left-wing parties.

From this point of view, the action taken by Brazil at the G20 since the return to power of Lula and the Workers' Party is particularly interesting. In 2024, the Brazilian government used its G20 presidency to promote the idea of a minimum tax for billionaires on an international level. The initiative was not adopted, but it did receive support from a large portion of G20 members, and the fact that it was suggested by a Southern country and not by a Western country is interesting in itself. When I proposed the implementation of a global minimum wealth tax in *Le Capital au XXIᵉ Siècle* (Capital in the twenty-first century) in 2013, I could not have imagined that such a measure would be the subject of an official discussion at the G20, especially as a Brazilian initiative. This shows that history is never written in advance and it will continue to surprise the skeptics, as the social-democratic movement did in its time.

What is certain is that if the Western countries refuse to face up to their historical responsibilities and share the wealth, then they expose themselves to more and more hostile reactions. Under the aegis of China and Russia, the BRICS group (Brazil, Russia, India, China, South Africa, Indonesia, Iran, Egypt, Ethiopia, and the United Arab Emirates) is in the process of integrating more countries, which could quickly lead to growing pressure on Western countries, which in turn would be well advised to reconsider their positions.[25]

Socialism, Liberalism, Nationalism: Three-Pillar Democracy

It is time to bring this text to a close and leave the reader to browse through the columns presented in the rest of this book, which include all my monthly op-eds published in *Le Monde* from February 2021 to January 2025, without any modifications or rewrites. These texts represent a social scientist's imperfect attempt to leave his ivory tower and his thousand-page historical books and to get involved in public debate and more immediate current affairs, with all the risks that that brings.

The call to democratic and ecological socialism formulated here must not distract from the essential: I believe above

---

25 See "Taking the BRICS Seriously," November 14, 2023. See also "Responding to the Challenge of China with Democratic Socialism," July 13, 2021.

all in the virtues of disagreement and public deliberation, in electoral pluralism and democratic alternation. Put simply, I am convinced that electoral democracy needs a strong socialist pillar to function properly. Since the nineteenth century, the political conflict has organized itself around three major ideological families: socialism, liberalism, and nationalism. Liberalism relies on private ownership and the domestic and international markets to promote individual emancipation and industrial development, sometimes with some economic success, but also considerable social damage. Nationalism responds to the ensuing social crisis by valorizing the nation and local and ethnonational solidarity, while socialism tries, not without difficulty, to promote an alternative socio-economic system founded on the sharing of power and ownership and universalist emancipation through education.

Each of these three main branches plays an indispensable role in bringing to the table of democratic deliberation the reasoning and social experiences that the other blocs need and must take into account. Throughout the twentieth century, the socialist pillar played a fundamental role: Not only was the social-democratic program founded on the social state applied, but its broad outlines gradually became consensual to the point that today, it is a part of the common democratic foundation. Nobody in the other blocs is really thinking about going back and removing social security, universal health coverage, or free education. What we have sometimes called "neoliberalism" since the 1980s and 1990s actually has nothing to do with pre-1914 liberalism: At the start of the twenty-first century, neoliberalism in Europe is aiming to stabilize the compulsory levies at 40–50% of national income (and therefore to leave the field open to lucrative logic and market forces to respond to the growing educational, health, and environmental needs), and not to reduce the compulsory levies to less than

10% of national income, which is very different. In the same way, nationalist Europeans of the 2010s and 2020s intend to defend territories forgotten by big-city globalists and to promote the interests of the "lower middle classes" caught between the racialized social security recipients and the hypocritical privileged. This vision can clearly be contested (it is, of course, much easier to target the poorest than the richest, but it is not a given that this solves the problems any more than ethnocentrism does), but it also expresses certain realities linked to the new geosocial class structure, which is characteristic of an advanced social state struggling with unbridled globalization.[26] In any case, the nationalists' program is not to remove the social state and return to the pre-1914 liberal state.

To summarize, in the twentieth century, the social-democratic revolution allowed the social state to become part of the common democratic foundation. We must proceed in the same way in the twenty-first century. Democratic and ecological socialism, based on a program of gradual decommodification of the economy (education, healthcare, energy, transportation, housing, etc.) and of the increasing socialization of wealth, must gradually enter the common foundation, as it will show its ability in the face of global social and environmental challenges in a much more convincing way than the market and capitalist logics. In the same way as in the twentieth century, these transformations will not simply be achieved by a peaceful process of collective and democratic deliberation. They will also feature moments of strong tensions, undoubtedly environmental catastrophes, and probably high-intensity geopolitical crises. These transformations will also require collective organizations capable of providing political outlets for

---

[26] See Cagé and Piketty, *Une histoire du conflit politique*.

crises and catastrophes and of structuring mobilizations, power relations, and social struggles. In the same way as all the major institutional transformations in the past, the implementation of twenty-first-century social democracy and democratic and ecological socialism will also require major legal and constitutional changes to allow for the redistribution of wealth that is indispensable for facing the environmental challenges and preserving the planet's habitability. This will almost inevitably involve crises and moments of tension, like almost all major legal and constitutional changes over the last two centuries.[27]

However, public deliberation and the reasoned acceptance of disagreement, within the framework of electoral and parliamentary democracy, will also play a central and irreplaceable role in these developments. The intellectual battle is at the heart of the political battle. Without a strong socialist and egalitarian pillar, the political conflict will too often be reduced to a false and deadly conflict between the nationalist and liberal elites. Social and economic questions belong to all citizens, and it is by overturning relations of knowledge and power that we can resume the march toward equality. I hope that the texts in this work will contribute to this vast collective undertaking.

October 2024

---

27 See "Can We Trust Constitutional Judges?," April 11, 2023. See also "The Fall of the U.S. Idol," January 12, 2021; "A Queen with No Lord?," September 13, 2022; "When the German Left Was Expropriating Princes," March 19, 2024.

Time for Social Justice
February 16, 2021

As the pandemic crisis fuels the demand for social justice more than ever, a new investigation by a consortium of international media (including *Le Monde*) has just revealed the financial turpitudes of Luxembourg, a tax haven nestled in the heart of Europe. There is an urgent need to get out of these contradictions and to launch a profound transformation of the economic system in the direction of justice and redistribution.

Let's start with the most immediate. The first priority should be social, wage, and ecological recovery. The Covid crisis has brought to light low pay in many key sectors. The CFDT, a union that is considered centrist, called in January for an immediate 15% increase in all low- and middle-wage workers in the medico-social sector. The same should be done in education, health, and all low-wage sectors.

It is also a time to radically accelerate the pace of thermal renovations of buildings, to create jobs on a massive scale in the environment and renewable energy sectors, and to extend minimum income systems to young people and students. Where should we stop in public recovery? The answer is simple: As long as inflation is near zero and interest rates are zero, we must continue. If and when inflation will sustainably return to a significant level (say, 3%–4% per year for two consecutive years), then it will be time to ease the pace.

The second step is that the highest private wealth will, of course, have to be used at some point to finance social recovery and reduce public debt.

This will require an increased effort at financial transparency. The OpenLux survey has shown that the register of the actual beneficiaries of companies (i.e., genuine owners, and not simply "shell" companies acting as fronts) made public by Luxembourg as a result of a European Union obligation, and which we are still waiting for France to put online, unfortunately has many flaws. The same is true of the system of automatic exchange of banking information set up by the OECD.

In general, all this new information is useful, but on condition that it is actually used by the tax authorities to actively involve the wealthy people who have previously evaded tax. Above all, it is essential that governments provide indicators that allow everyone to check how far this is moving toward a fairer tax system.

In concrete terms, tax authorities must publish detailed information each year on the taxes paid and the overlaps made concerning the different categories of taxpayers concerned. As with the registers of actual beneficiaries, information should ideally be nominal, especially for the largest companies and the highest fortunes.

If it is decided that this is not desirable, then at the very least, the published statistical information should make clear the taxes paid by people in very high wealth brackets: fortunes between 1 and 10 million euros, between 10 and 100 million euros, between 100 million and 1 billion euros, and so on. Model tables have been proposed by the World Inequality Lab (see table 1), and they could naturally be discussed and improved.

The general idea is simple: Billionaires are everywhere in magazines, and it's time for them to appear in tax statistics.

According to *Challenges*, the top 500 French fortunes rose from 210 to 730 billion euros between 2010 and 2020 (from 10% to 30% of GDP). How did their taxes change during this period? No one knows. If governments have really made the dramatic progress in terms of transparency that they claim to have made in recent years, then it is time for them to prove this by making this type of information public.

If we extend the focal point of the first 500 fortunes (beyond 150 million euros of individual assets according to *Challenges*) to the 500,000 highest assets (about 1% of the adult population, with assets in excess of 1.8 million euros according to the World Inequality Database), then the total fortunes concerned reach 2,500 billion euros (nearly 120% of GDP), thus increasing the tax stakes.

To go beyond the prevailing conservatism, it is also urgent to go back to history. After the Second World War, when public debt had reached levels higher than today, most countries introduced exceptional levies on the highest private assets. This is particularly the case in Germany with the *Lastenausgleich* system (or "burden-sharing" system, which was the subject of an excellent historical study by Michael Hughes) adopted by the Christian Democrat majority in 1952. With a levy of up to 50% on the highest financial and real estate holdings, payable over 30 years, this system yielded 60% of GDP to the state, at a time when billionaires were much less prosperous than today.

Combined with the monetary reform of 1948 and the cancellation of the external debt in 1953, this system allowed Germany to get rid of its public debt without resorting to inflation (which had done so much harm to the country in the 1920s) and by relying on a credible social justice objective.

It is high time to go back to the roots of what made the success of post-war European reconstruction.

Table 1. Tracking Progress toward Global Financial Transparency and Tax Justice: Public Statistics to Be Published by National Tax Administrations

Summary: In order to track inequality, progress toward global financial transparency, and tax justice, all countries should commit to publish on an annual basis the following tables. This applies in particular to the countries participating in the various international discussion groups on these issues, in particular those coordinated by OECD on CRS (Common Reporting Standards on cross-border financial assets) and BEPS (Base Erosion and Profit Shifting on corporate taxation).

Net wealth: Total assets (real estate, business, financial, etc.), net of debt. For country residents, all domestic and foreign assets should be included. For non-residents, all domestic assets should be included (in particular real estate assets located in the country), as well as all financial assets related to firms and economic activities conducted in the country). To the extent possible, their foreign assets should also be included.

Table 1A: Number of Individuals, Wealth and Taxes Paid by Wealth Bracket

Net wealth bracket (€)	Number of individuals	Incl. number of residents	Incl. number of non-residents	Total net wealth	Incl. residents	Incl. non-residents	Total wealth taxes	Wealth taxes — Incl. wealth and property tax	Wealth taxes — Incl. capital gains tax	Wealth taxes — Incl. inheritance & estate tax	Total income taxes	Income taxes — Incl. personal income tax	Income taxes — Incl. corp. income taxes
0–10k													
10k–100k													
100k–1m													
1m–10m													
10m–100m													
100m–1bn													
1bn–5bn													
5bn–10bn													
10bn+													

Table 1B: Wealth and Income Composition by Wealth Bracket

Net wealth bracket (€)	Wealth										Income		
	Total wealth	Incl. currency & deposits	Incl. bonds & loans	Incl. equities & fund shares	Incl. pension funds & life insur.	Inc. real estate	Incl. business & other non-fin. assets	Incl. debt	Incl. total domestic assets	Incl. total foreign assets	Total income	Incl. capital income	Incl. labor income
0–10k													
10k–100k													
100k–1m													
1m–10m													
10m–100m													
100m–1bn													
1bn–5bn													
5bn–10bn													
10bn+													

Source: L. Chancel, "Measuring Progress towards Tax Justice," World Inequality Lab, 2019.

Combatting Discrimination, Measuring Racism
March 16, 2021

As the trial of George Floyd's killer opens in the United States, identity conflicts are festering in Europe and France. Instead of fighting discrimination, the government has embarked on a course of pursuing the far right and hunting down social scientists. This is all the more regrettable because there is an urgent need to set up a genuine French and European model to combat discrimination. A model which would embrace the reality of racism and ensure the means to measure and correct it, while placing the fight against discrimination within the broader framework of a social policy with a universalist agenda.

Let me start with the question of measuring racism. Numerous research studies have demonstrated the reality of racism, but we lack a real Observatory of Discrimination that objectifies the facts and ensures annual monitoring. The Defender of Rights, an institution which replaced the Halde (Haute Autorité de lutte contre les discriminations et pour l'égalité) in 2011, stresses in its reports the extent of discrimination in employment or housing, but still does not have the means to monitor it systematically.

Combatting Discrimination, Measuring Racism

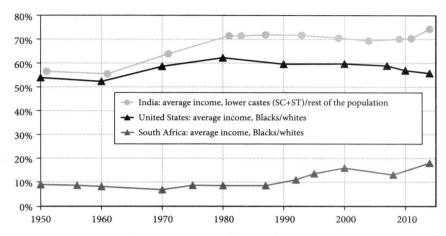

Figure 9. Discrimination and inequality in comparative perspective. The ratio between the average income of lower castes in India (scheduled castes and tribes, SC+ST, ancient discriminated groups of untouchables, and aboriginal tribes) and that of the rest of the population rose from 57% in 1950 to 74% in 2014. The ratio between the average income of Blacks and whites over the same period rose from 54% to 56% in the United States, and from 9% to 18% in South Africa.

Sources and series: piketty.pse.ens.fr/ideology, figure 8.6.

For example, in a 2014 study, researchers sent fake CVs to employers in response to some 6,231 job advertisements and observed the response rates in the form of a job interview offer. If the name is Muslim-sounding, the response rate is divided by four. Jewish-sounding names are also discriminated against, although less massively. The problem is that this study has not been repeated, so no one knows whether the situation has improved or worsened since 2014.

There is an urgent need for an official Observatory to say how these indicators are evolving annually. This requires large-scale testing campaigns that allow for reliable comparisons over time and between regions and sectors of activity. It is also essential to be able to say to what extent discrimination is concentrated among a fraction of employers. Like anti-Semitism or homophobia, Islamophobia is not inevitable and can be overcome. Discussion must also take place on the term to be used: some prefer to speak of anti-Muslimism or anti-Muslim discrimination. Why not, but on condition that this does not prevent progress on the substance.

The Discrimination Observatory, which could be placed under the authority of the Defender of Rights, should also ensure the annual monitoring of discrimination within companies (salaries, promotions, training, etc.). To do this, questions on the parents' country of birth should be introduced into the census surveys (which are carried out amongst 14% of the population each year). This has long been the case in many public surveys (employment surveys, FQP [Formation, Qualification Professionale] or Trajectories and Origins). But their frequency and size are insufficient to make breakdowns by region, sector, and size of company, which would be possible with censuses on an anonymous basis. Without such indicators, it is impossible to fight discrimination effectively.

The important point is that all this can be done without introducing ethno-racial categories such as those used in the United States and United Kingdom. The problem is not so much that this is constitutionally prohibited (the Constitution was amended in 1999 and 2008 to allow for gender parity), but rather that such categories would run the risk of rigidifying multiple and mixed identities, with no clear evidence of their effectiveness in combatting discrimination. Since their introduction in 1991 in the British censuses, there is no indication that discrimination has decreased in the United Kingdom compared to other countries.

There is also considerable confusion in individual responses: Between a quarter and a half of people born in Turkey,

Egypt, or the Maghreb classify themselves as "White" (a category with which they identify better than "Black/Caribbean" or "Indian/Pakistani"), while others classify themselves as "Asian" or "Arab" (a category introduced in 2011, but which did not appeal to all of the people targeted).

If no European country has replicated this experience of identity assignment, this does not necessarily mean that no one cares about discrimination in France, Germany, Sweden, or Italy. The introduction of objective questions on the country of birth of parents in censuses poses fewer difficulties and would allow for real progress, with annual testing and real political monitoring.

More generally, positive discrimination policies developed on the basis of ethno-racial categories in the United States or the United Kingdom, castes in India or territories in France are often very hypocritical. This gives the authorities a clear conscience at little cost, while the issue of financing the public services that are essential to breaking the cycle of inequality is often forgotten.

As Asma Benhenda has shown, the average salary per teacher in France increases sharply with the proportion of socially advantaged pupils in the school. In other words, the meager bonuses distributed in priority education zones are in no way sufficient to compensate for the over-representation of temporary and inexperienced teachers. When we create 1,000 places in "talented" prepas (preparatory classes) without increasing the resources allocated to the millions of disadvantaged students in university tracks, we are only reinforcing a hyper-inegalitarian education system.

We must have the means to measure and vigorously fight against discrimination, but we must also and above all support universal social policies, without which the march toward equality will remain wishful thinking.

Rights for Poor Countries
April 13, 2021

The Covid-19 crisis, the most serious global health crisis in a century, forces us to fundamentally rethink the notion of international solidarity. Beyond the right to produce vaccines and medical equipment, it is the whole question of the right of poor countries to develop and to receive part of the tax revenues of the world's multinationals and billionaires that must be asked. We need to move beyond the neocolonial notion of international aid, paid at the whim of rich countries and under their control, and finally move toward a logic of rights.

Let's start with vaccines. Some argue (unwisely) that there would be no point in lifting patent ownership rights because poor countries would be unable to produce the precious doses. This is not true. India and South Africa have significant vaccine production capacity, which could be expanded, and medical supplies can be produced almost anywhere. It is not simply to pass the time that these two countries have taken the lead in a coalition of some 100 countries to demand that the WTO agree to the exceptional lifting of these patent ownership rights. By opposing this, the rich countries have not only left the field open to China and Russia: They have missed a great opportunity to change times and show that their conception of multilateralism is not one-sided. Let's hope they reverse course soon.

But beyond this right to produce, it is the entire international economic system that must be rethought in terms of the rights of poor countries to develop and to no longer allow themselves to be plundered by the richest. In particular, the debate on the reform of international taxation cannot be reduced to a discussion between rich countries aimed at sharing the profits currently located in tax havens. This is the problem with the plans being discussed at the OECD. It is envisaged that multinationals will make a single declaration of their profits at the global level, which in itself is an excellent thing. But when it comes to allocating this tax base between countries, it is planned to use a variety of criteria (wage bill and sales in different territories), which in practice will result in rich countries receiving over 95% of the reallocated profits, leaving only crumbs for poor countries. The only way to avoid this predicted disaster is to finally include the poor countries around the table and to distribute the profits in question according to the population (at least in part).

This debate must also be seen in the broader perspective of a progressive tax on the highest incomes and wealth, not just a minimum tax on the profits of multinationals. In concrete terms, the 21% minimum rate proposed by the Biden administration is a significant step forward, especially as the United States plans to apply it immediately, without waiting for an international agreement. In other words, subsidiaries of US multinationals based in Ireland (where the rate is 12%) will immediately pay an additional 9% tax to Washington. France and Europe, which continue to defend a minimum rate of 12%, which would not change anything, seem completely overwhelmed by events. But this system of minimum tax on multinationals is nonetheless highly insufficient if it is not part of a more ambitious perspective aimed at restoring tax progressiveness at the individual level. The OECD suggests amounts of less than 100 billion, or less than 0.1% of global GDP (about 100 trillion euros).

By comparison, a global tax of 2% on fortunes over €10 million would raise ten times as much: €1,000 billion per year, or 1% of global GDP, which could be allocated to each country in proportion to its population. Setting the threshold at €2 million would raise 2% of global GDP, or even 5% with a highly progressive scale for billionaires. If we stick to the least ambitious option, this would be more than enough to entirely replace all current international public aid, which represents less than 0.2% of world GDP (and barely 0.03% for emergency humanitarian aid, as Pierre Micheletti of Action Against Hunger recently pointed out).

Why should every country be entitled to a share of the revenues collected from the world's multinationals and billionaires? Firstly, because every human being should have equal minimum rights to health, education, and development. Secondly, because the prosperity of the rich countries would not exist without the poor countries: Western enrichment has always been based on the international division of labor and the unbridled exploitation of the world's natural and human resources. Of course, rich countries could continue to fund their development agencies if they so wished. But this would be in addition to the irrevocable right of poor countries to develop and build their states.

To prevent money from being misused, the tracking of ill-gotten wealth, whether from Africa, Lebanon, or any other country, should also be generalized. The system of uncontrolled capital flows and financial opacity imposed by the North since the 1980s has done much to undermine the fragile process of state-building in the South, and it is time to end it.

Lastly, there is nothing to prevent each rich country from starting now to allocate a fraction of the taxes levied on multinationals and billionaires to poor countries. It is time to pick up the new wind coming from the United States and carry it in the direction of a sovereignism driven by universalist objectives.

From Basic Income to Inheritance for All

May 18, 2021

The Covid crisis is forcing us to rethink the tools of redistribution and solidarity. Proposals are springing up everywhere: basic income, job guarantee, inheritance for all. Let's say it straight away: These proposals are complementary and not substitutable. In the long run, they must all be implemented, in stages and in this order.

Let's start with basic income. Such a system is dramatically lacking today, especially in the South, where the incomes of the working poor have collapsed and containment rules are unenforceable in the absence of a minimum income. Opposition parties had proposed introducing a basic income in India in the 2019 elections, but the ruling nationalist-conservatives in Delhi are still dragging their feet.

In Europe, various forms of minimum income exist in most countries, but with many shortcomings. In particular, there is an urgent need to extend access to them to younger people and students (this has already long been the case in Denmark), and especially to people without a fixed address or bank account, who often face insurmountable obstacles.

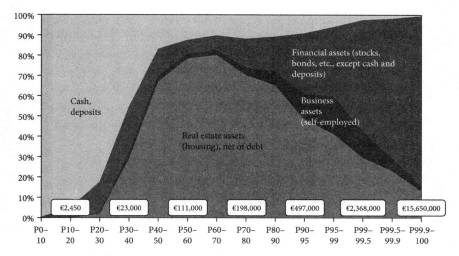

Figure 10. Composition of property, France, 2015. In France in 2015 (as in most countries where data are available), small fortunes consist primarily of cash and bank deposits, medium fortunes of real estate, and large fortunes of financial assets (mainly stocks).

Note: The distribution shown here is per adult wealth (wealth of couples divided by 2).

Sources and series: piketty.pse.ens.fr/ideology, figure 11.17.

It is worth noting in passing the importance of the current discussions on central bank digital currencies, which should ideally lead to the creation of a genuine public banking service, free and accessible to all, the antithesis of the systems dreamed up by private operators (whether decentralized and polluting, like bitcoin, or centralized and inegalitarian, like the projects of Facebook or private banks).

It is also essential to extend the basic income to include low-paid workers, with a system of automatic payment on pay slips and bank accounts, without people having to ask for it, linked to the progressive tax system (also deducted at source).

Basic income is an essential but insufficient tool. In particular, its amount is always extremely modest. Depending on the proposal, it is generally between half and three quarters of the full-time minimum wage, so that, by construction it can only be a partial tool in the fight against inequality. For this reason, it is preferable to speak of a basic income than of a universal income (a notion that promises more than this minimalist reality).

A more ambitious tool that could be implemented as a complement to the basic income is the job guarantee system recently proposed in the framework of the discussions on the Green New Deal.[1] The idea is to offer all those who want it a full-time job at a minimum wage set at a decent level ($15 per hour in the United States). Funding would be provided by the federal government, and jobs would be offered by public employment agencies in the public and voluntary sectors (municipalities, communities, nonprofit organizations).

Placed under the dual patronage of the Economic Bill of Rights proclaimed by Roosevelt in 1944 and the March for Jobs and Freedom organized by Martin Luther King in 1963, such a system could make a powerful contribution to the process of de-commodification and collective redefinition of needs, particularly in the areas of personal services, energy transition, and renovation of buildings. It also enables, at a limited cost (1% of GDP in Tcherneva's proposal), to bring back into employment all those who were deprived of a job during recessions and thus avoid irremediable social damage.

Finally, the last device that could complete the package, in addition to the basic income, the job guarantee, and the set

---

[1] See in particular Pavlina Tcherneva, *The Case for a Job Guarantee* (Polity Press, 2020).

of rights associated with the most extensive social state possible (free education and health, highly redistributive pensions and unemployment benefits, trade union rights, etc.), is a system of inheritance for all.

When we study inequality in the long term, the most striking thing is the persistence of a hyper-concentration of property. The poorest 50% have hardly ever owned anything: 5% of total wealth in France today, compared to 55% for the richest 10%. The idea that we just have to wait for wealth to spread doesn't make much sense: If that were the case, we would have seen it long ago.

The simplest solution is a redistribution of inheritance allowing the whole population to receive a minimal inheritance, which, to fix ideas, could be of the order of 120,000 euros (i.e., 60% of the average wealth per adult). Paid to all at 25 years of age, it would be financed by a mix of progressive wealth and inheritance taxes yielding 5% of national income (a significant amount but one that could be considered in the long term). Those who currently inherit nothing would receive 120,000 euros, while those who inherit a million euros would receive 600,000 euros after taxation and endowment. We are therefore still far from equality of opportunity, a principle often defended at a theoretical level, but which immediately puts the privileged classes on their guard whenever one envisages a beginning of concrete application. Some will want to put constraints on its use; why not, provided they apply to all inheritances.

Inheritance for all aims to increase the bargaining power of those who have nothing, to enable them to refuse certain jobs, to acquire housing, or to embark on a personal project. This freedom is frightening for employers and property owners, as it would make workers less docile, but liberating for all others. We are only just emerging from a long period of enforced confinement. All the more reason to start thinking and hoping again.

The G7 Legalizes the Right to Defraud

June 15, 2021

Last weekend, the G7 ministers announced their intention to apply a minimum tax rate of 15% on the offshore profits of multinationals. Let us be clear: If we leave it at that, it is nothing more and nothing less than the formalization of a real license to defraud for the most powerful players. For small and medium-sized enterprises as well as for the working and middle classes, it is impossible to create a subsidiary to relocate its profits or income to a tax haven. For all these taxpayers, there is no choice but to pay ordinary tax. However, if we add up taxes on income and profits and social security contributions, both employees and the small and medium-sized self-employed find themselves paying rates in all the G7 countries well above 15%: at least 20–30%, and often 40–50%, or even more.

The G7 announcement is all the more inappropriate because the ProPublica website has just published a vast survey confirming what the researchers had already shown: American billionaires pay almost no income tax compared to the extent of their enrichment and what the rest of the population pays. In practice, the corporate tax is often the final tax paid by the richest (when they pay it). Profits accumulate in companies or ad hoc structures (trusts, holding companies, etc.), which finance most of the lifestyle of the people in question (private jets, bank cards, etc.), almost without any control. By legalizing the fact that multinationals will be able to continue to locate their profits in tax havens at leisure, with a tax rate of 15% as the only tax, the G7 is formalizing entry into a world where oligarchs pay structurally less tax than the rest of the population.

How can we break this deadlock? Firstly, by setting a minimum rate higher than 15%, which each country can do right now. As the European Tax Observatory has shown, France could apply a minimum rate of 25% to multinationals, which would bring it 26 billion euros per year, equivalent to almost 10% of health spending. With a rate of 15%, only slightly higher than the rate applied in Ireland (12.5%), which makes the measure harmless, the revenue would be barely 4 billion. Part of the 26 billion could be used to improve the financing of the hospitals, schools, and the energy transition; another part to lighten the tax burden on the self-employed and less prosperous employees. What is certain is that it is illusory to expect European unanimity on such a decision. Only unilateral action, ideally with the support of a few countries, can unblock the situation. Ireland or Luxembourg will undoubtedly lodge a complaint with the European Court of Justice, arguing that the principles of absolute free movement of capital (without any fiscal, social, or environmental compensation) defined 30 years ago do not provide for such action. It is difficult to say how the ECJ will decide, but if necessary these rules will have to be denounced and rewritten.

Furthermore, it is urgent to remember that the tax on corporate profits cannot be the final tax for shareholders or managers of companies. It must once again become what it should never have ceased to be, namely an advance payment within the framework of an integrated tax system with progressive income tax at the individual level. The G7 discussions must be explicitly framed within this framework. In theory, rich countries are believed to have implemented systems for the automatic transmission of international banking information on cross-border holdings and individual financial income in recent years. Why, then, do they not publish reliable indicators to measure progress? Specifically, the G7 countries should publish detailed information each year showing the taxes paid by people belonging to very high income and wealth groups (fortunes between 1 and 10 million euros, between 10 and 100 million euros, between 100 million euros and 1 billion euros, and so on). Judging by the ProPublica survey, one would probably realize that the wealthiest do not pay much, given the possibilities of downward manipulation of their individual tax income, and that only a progressive wealth tax would make it possible to tax them significantly and in relation to their wealth. In any case, rather than waiting for the next revelations, all governments should immediately make public the amount of taxes paid by their billionaires and millionaires, especially in France.

Last but not least, this discussion must finally be opened up to the countries of the South. The mechanism envisaged by the G7, according to which each country is responsible for charging a minimum tax to its own multinationals, is only acceptable if it is an advance payment within a broader system of revenue distribution. The G7 raises the possibility that part of the profits above a certain break-even point (more than 10% per year of the capital invested) will be distributed according to sales in the different countries. But this system will only cover tiny sums and will essentially be reduced to redistribution between the countries of the North. If the latter really want to take up the Chinese challenge, improve their degraded image, and above all give the South a chance to develop and build viable states, it is urgent that poor countries have a significant share of the revenues of the multinationals and billionaires of the planet.

Responding to the Challenge of China with Democratic Socialism
July 13, 2021

As the Chinese Communist Party (CCP) commemorates its 100th anniversary, Western countries are still struggling to define their attitude toward the Beijing regime. Let me say it straight away: The right answer lies in ending Western arrogance and promoting a new emancipatory and egalitarian horizon on a global scale, a new form of democratic and participatory, ecological and post-colonial socialism. If they stick to their usual lecturing posture and a dated hyper-capitalist model, Western countries may find it extremely difficult to meet the Chinese challenge.

Authoritarian and oppressive, the Beijing regime certainly has many weaknesses. According to the *Global Times,* the official daily newspaper, Chinese-style democracy is superior to the Western-style electoral supermarket, because it entrusts the country's destinies to a motivated and determined avant-garde, both selected and representative of society (the CCP has 90 million members, 10% of the population), and ultimately more deeply involved in the service of the general

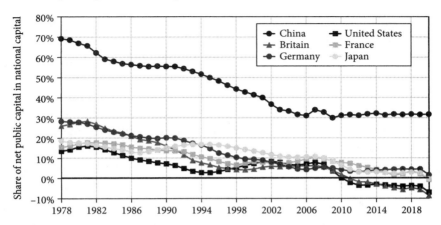

Figure 11. The fall of public property, 1978–2020. The share of public capital (public assets net of debt, all government levels and asset categories combined: companies, buildings, land, financial assets, etc.) in national capital (i.e., the sum of public and private capital) was about 70% in China in 1978, and it has stabilized at around 30% since the mid-2000s. This share was around 15%–30% in capitalist countries in the 1970s and is near zero or negative in 2020.

Sources and series: piketty.pse.ens.fr/equality, figure 39.

interest than the average Western voter, versatile and impressionable.

In practice, however, the regime is becoming more and more like a perfect digital dictatorship, so perfect that no one wants to look like it. The model of deliberation within the party is all the less convincing because it leaves no trace outside, while on the contrary everyone can see more and more clearly the establishment of widespread surveillance on social networks, the repression of dissidents and minorities, the brutalization of the electoral process in Hong Kong, threats to electoral democracy in Taiwan. The ability of such a regime to appeal to the views of other countries (and not just their leaders) seems limited. We must add the sharp rise in inequality, the acceleration in aging, the extreme opacity that characterizes the distribution of wealth, the feeling of social injustice that results from this, and that cannot be permanently appeased by a few marginalizations.

Despite these weaknesses, the regime has strong assets. When climatic disasters occur, it will have no difficulty in stigmatizing the responsibilities of the former powers, which despite their limited population (about 15% of the world's population for the United States, Canada, Europe, Russia, Japan taken together) account for nearly 80% of cumulative carbon emissions since the beginning of the industrial era.

More generally, China does not hesitate to recall that it industrialized without resorting to slavery and colonialism, of which it itself has borne the brunt. This allows it to score points in the face of what is perceived throughout the world as the eternal arrogance of Western countries, always quick to give lessons to the whole world in terms of justice and democracy, while showing themselves incapable of facing the inequalities and discriminations that undermine them and by making a pact with all the potentates and oligarchs who benefit from them.

On the economic and financial level, the Chinese state has considerable assets, much greater than its debts, which gives it the means for an ambitious policy, both domestically and internationally, especially regarding investments in infrastructure and in the energy transition. The public authorities currently hold 30% of everything there is to own in China (10% of real estate, 50% of companies), which corresponds to a mixed economy structure that is in many ways comparable to that found in the West during the period of prosperity 1950–1980 (known in France as the Trente Glorieuses, or "Glorious Thirty").

Conversely, it is striking to note the extent to which the main Western states all find themselves in the early 2020s with

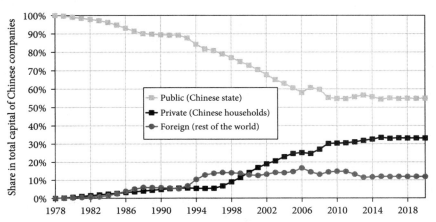

Figure 12. Ownership of Chinese firms, 1978–2020. The Chinese state (all government levels combined) in 2017 owned about 55% of total capital of Chinese firms (both listed and unlisted, of all sizes and all sectors) versus 33% for Chinese households and 12% for foreign investors. The foreign share has diminished since 2003, and that of Chinese households increased, while that of the Chinese state stabilized at around 55%.

Sources and series: piketty.pse.ens.fr/equality, figure 40.

almost zero or negative patrimonial positions. As a result of not having balanced their public accounts (which would have required a larger contribution from the richest taxpayers), these countries have accumulated public debts, while selling an increasing proportion of their public assets, so much so that the former ended up slightly overtaking the latter. Let me be very clear: Rich countries are rich, in the sense that private wealth has never been so high; it is only their states that are poor. If they persist in this direction, they could end up with an increasingly negative public patrimony, which would correspond to a situation where the holders of debt securities own not only the equivalent of all public assets (buildings, schools, hospitals, infrastructure, etc.), but also a drawing right on an increasing share of future taxpayers' taxes. On the other hand, it would be quite possible, as was done in the post-war period, to reduce public debt in an accelerated manner by draining the highest private assets, and thus to restore room for maneuver to the public authorities.

It is at this price that we will return to an ambitious policy of investment in education, health, the environment, and development. There is also an urgent need to lift the rights on vaccines, to share the revenues of multinationals with the countries of the South, and to put digital platforms at the service of the general interest. More generally, there is a need to promote a new economic model based on the sharing of knowledge and power at all levels, in business as well as in international organizations. Neoliberalism, by handing over power to the richest and weakening public power, in the North as well as in the South, has in reality only strengthened the Chinese model, as indeed has the pathetic Trumpist or Modist neo-nationalism. It is time to move on.

Emerging from September 11
September 14, 2021

Twenty years ago, the World Trade Center towers were struck by airplanes. The worst attack in history was to lead the United States and some of its allies into a global war against terrorism and the "axis of evil." For the US neo-conservatives, the attack was proof of the theses put forward by Samuel Huntington in 1996: The "clash of civilizations" was becoming the new way of interpreting the world. This publication was their oft-quoted favorite, just as the works published by Milton Friedman in the 1960s and 1970s were those of the Reaganites in the 1980s.

Unfortunately, we now know that the US desire for revenge and the resulting brutalization of entire regions and societies has only exacerbated identity-based conflicts. The invasion of Iraq in 2003, with its state-sponsored lies about weapons of mass destruction, only served to undermine the credibility of the "democracies." With the images of US soldiers holding prisoners on leashes at Abu Ghraib, there was no need for recruiting agents for the jihadists. The unrestrained use of force, the arrogance of the US Army, and the enormous civilian losses among the Iraqi population (at least 100,000 deaths acknowledged) did the rest and contributed powerfully to the decomposition of the Iraqi-Syrian territory and to the

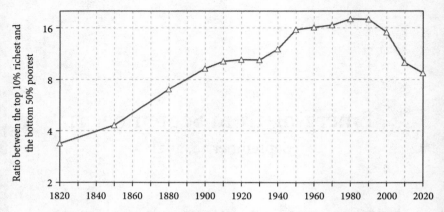

Figure 13. Income gaps between countries, 1820–2020: the slow exit from colonialism. Income gaps between countries, as measured by the ratio between the average income of the top 10% of the world population living in the richest countries and the bottom 50% of the population living in the poorest countries, increased significantly between 1820 and 1960–1980 before beginning a period of reduction.

Note: For the computation of this ratio, the population of overlapping countries has been divided between deciles as if they were multiple countries.

Sources and series: piketty.pse.ens.fr/equality, figure 36.

rise of the Islamic State. The terrible failure in Afghanistan, with the return of the Taliban to power in August 2021, after twenty years of Western occupation, symbolically concludes this sad sequence.

To truly emerge from September 11, a new reading of the world is necessary: It is time to abandon the notion of a "war of civilizations" and replace it with that of co-development and global justice. This requires explicit and verifiable objectives of shared prosperity and the definition of a new economic model, sustainable and equitable, in which each region of the planet can find its place. Everyone now agrees that the military occupation of a country only strengthens the most radical and reactionary segments and can do no good. The risk is that the military-authoritarian vision will be replaced by a form of isolationist withdrawal and economic illusion: The free movement of goods and capital will be enough to spread wealth. This would be to forget the highly hierarchical nature of the global economic system and the fact that each country does not fight on equal terms.

From this point of view, a first opportunity was missed in 2021: The discussions on the reform of the taxation of multinationals essentially boiled down to revenue sharing between rich countries. However, it is urgent that all countries, in the North as well as in the South, receive a share of the revenues from the most prosperous global players (multinationals and billionaires), according to their population. Firstly, because every human being should have an equal minimum right to development, health, and education; and secondly, because the prosperity of rich countries would not exist without poor countries. Western enrichment yesterday or Chinese enrichment today has always been based on the international division of labor and the unbridled exploitation of the world's human and natural resources. When refugees appear on the other side of the world, Westerners like to explain that it is up to the neighboring countries to take care of them, however poor they may be. But when there is uranium or copper to be mined, Western companies are always there first, no matter how far away.

If we accept the principle of revenue sharing between all countries, then we obviously need to talk about the allocation criteria and the rules to be respected in order to be entitled to it. This would be an opportunity to define precise and demanding rules regarding respect for human rights, and in particular the rights of women and minorities, which would apply to the Taliban as well as to all countries wishing to benefit from the windfall. To prevent the money from being misused, the tracking of ill-gotten gains should also be generalized and there should be full transparency on excessive enrichment, whether in the public or private sector, in the South or in the North. The central point is that the criteria must be defined in a neutral and universal way and applied everywhere in the same way, in Afghanistan as in Saudi Arabia and the petro-monarchies, in Paris as in London or Monaco. Western countries must stop using the corruption argument at every turn to deny the South's right to self-governance and development, while at the same time pandering to the despots and oligarchs who benefit them. The time for unconditional free trade is over: Further trade must depend on objective social and environmental indicators.

It is certainly understandable that Biden wants to turn the page on the war of civilizations as soon as possible. For the United States, the threat is no longer Islamist: It is Chinese, and above all it is internal, with social and racial fractures threatening the country and its institutions with a new quasi-civil war. But the fact is that the Chinese challenge, like the domestic social challenge, can only be solved by transforming the economic model. If nothing is proposed in this sense, then it is increasingly toward Beijing and Moscow that the poor countries and the peripheral and forgotten regions of the planet will turn to finance their development and maintain order. The way out of 9/11 should not lead to a new form of isolationism, but instead to a new climate of internationalism and universalist sovereignty.

"Pandora Papers": Maybe It Is Time to Take Action?
October 12, 2021

After the "LuxLeaks" in 2014, the "Panama Papers" in 2016, and the "Paradise Papers" in 2017, the revelations of the "Pandora Papers," resulting from a new leak of 12 million documents from offshore finance, show the extent to which the wealthiest continue to evade taxes. Contrary to what is sometimes claimed, there is no reliable indicator that the situation has improved over the last ten years.

Before the summer, ProPublica revealed that US billionaires pay almost no taxes compared to their wealth and what the rest of the population pays. According to *Challenges,* the top 500 French fortunes jumped from 210 billion euros in 2010 to more than 730 billion in 2020 (i.e., from 10% of GDP to 30% of GDP), and everything suggests that the taxes paid by these large fortunes (quite simple information, but which the public authorities still refuse to publish) were extremely low. Should we just wait for the next leaks, or is it not time for the media and citizens to formulate a platform for action and put pressure on governments to address the issue in a systemic way?

The basic problem is that at the beginning of the twenty-first century we continue to register and tax assets solely on the basis of real estate, using the methods and cadastres established at the beginning of the nineteenth century. If we do not provide ourselves with the means to change this state of affairs, then the scandals will continue, with the risk of a slow disintegration of our social and fiscal pact and the inexorable rise of every man for himself.

The important point is that registration and taxation of property have always been closely linked historically. Firstly, because registering one's property provides the owner with an advantage (that of benefiting from the protection of the legal system), and secondly, because only minimal taxation can make registration truly compulsory and systematic.

Moreover, the ownership of assets is also an indicator of the taxpaying capacity of individuals, which explains why the taxation of assets has always played a central role in modern tax systems, in addition to the taxation of the income stream (which can sometimes be manipulated downwards, especially for very rich asset holders, as ProPublica has shown).

By setting up a centralized register for all real estate assets, both housing and business property (farmland, shops, factories, etc.), the French Revolution also instituted at the same time a system of taxation based on asset transactions (transfer duties, still in force today) and above all on asset ownership (with the *taxe foncière*).

In France, as in the United States and in almost all rich countries, the *taxe foncière* or its Anglo-Saxon equivalent, the property tax, continues to represent the main tax on wealth (around 2% of GDP, approximately 40 billion euros in annual revenue in France). Conversely, the absence of such a system for registering and taxing housing and business assets explains the hypertrophy of the informal sector in many countries of the South and the subsequent difficulties in implementing income taxation.

The problem is that this system of registration and taxation of assets has hardly changed for two centuries, even though financial assets have become increasingly important. The result is an extremely unfair and unequal system. If you own a house or a business asset worth 300,000 euros, and you are in debt to the tune of 290,000 euros, then you will pay the same *taxe foncière* or property tax as someone who has inherited the same property and also has a financial portfolio of 3 million euros.

No principle, no economic reasoning, can justify such a violently regressive tax system (small assets holders de facto pay a structurally higher effective rate than larger ones), apart from the fact that it is assumed that it would be impossible to register financial assets. However, this is not a technical impossibility but a political choice: The choice was made to privatize the registration of financial securities (with central depositories under private law, such as Clearstream or Eurostream) and then to introduce the free movement of capital guaranteed by the States, without any prior tax coordination.

The "Pandora Papers" also remind us that the wealthiest people manage to avoid taxes on their real estate assets by transforming it into financial securities domiciled offshore, as shown by the case of the Blair family and their 7-million-euro house in London (400,000 euros in transfer duties avoided) or that of the villas held on the Côte d'Azur via shell companies by the Czech prime minister Babis (who is also suspected of embezzling European funds).

What should be done? The priority should be the establishment of a public financial register and a minimum taxation of all assets, if only to produce objective information about them. Each country can move immediately in this direction by requiring all companies holding or operating assets on its territory to reveal the identity of their owners and taxing them accordingly, transparently, and in the same way as ordinary taxpayers, no more and no less. By abandoning any ambition in terms of fiscal sovereignty and social justice, we only encourage the separatism of the richest and the withdrawal into ourselves. It is high time for action.

Can the French Presidential Election Be Saved?
November 16, 2021

With less than five months to go before the first round, what can we expect from the French presidential election scheduled for next April? The question can be asked at two levels: that of the 2022 election, and the broader question of the place of the presidential election in the French political system.

As far as the 2022 election is concerned, we have to admit that it is not off to a good start. Given the increasing tendency of the political landscape toward the extreme right-wing, an evolution to which Macronism in power is no stranger, it has become almost impossible to debate the major social and economic issues that will structure our common future.

To win the battle for emancipation, intelligence, and human capital, the central issue remains investment in education and training. Unfortunately, the latest figures from the 2022 finance law are clear: Public spending per student has fallen by 14% in France between 2008 and 2022 (–7% since 2017).

This is a monumental waste for the country and its youth. It is urgent that the candidates commit themselves to precise objectives allowing universities to finally have the same means as the selective courses and to develop the multidisciplinary courses and the levels of supervision that students need.

To meet the climate challenge, we know that efforts will have to be better distributed and that the wealthiest will have to pay more in taxation. Exempting the wealthiest from all taxation, even though they have tripled in number in France over the past 10 years, is economic stupidity and ideological blindness. This abandonment of any ambition in terms of fiscal sovereignty and social justice aggravates the separatism of the richest and feeds the headlong rush toward regalian and identity issues.

But whatever we do to ignore the primacy of social issues and inequalities, reality will rapidly return. In France, the poorest 50% have a carbon footprint of barely 5 tonnes per capita, compared to 25 tonnes for the richest 10% and 79 tonnes for the richest 1%. Solutions that consist in taxing everyone at the same rate, like the carbon tax at the beginning of the five-year term, make little sense and can never be accepted.

We could multiply the subjects: Local taxation must be rethought to allow the poorest municipalities and their inhabitants to have the same opportunities as others; the pension system must become universal and fair, with an emphasis on small and medium-sized pensions; a new sharing of power must be applied between employees and shareholders in the governance of companies; the fight against discrimination must become an assumed and measurable priority.

The candidates must also say whether they will be satisfied with the minimalist 15% rate on multinationals or whether they will commit to raising this rate unilaterally to 25%, as recommended by the European Tax Observatory, and to sharing the revenue with the countries of the South, which are severely affected by global warming and underdevelopment. Beyond these necessary unilateral decisions, it is urgent to propose to our European partners the setting up of a transnational Assembly that would allow for the adoption of common social, fiscal, budgetary, and environmental measures by a majority. This can probably only be done initially with a few countries. The issue is nonetheless crucial: The laborious debates on the recovery plan have shown the limits of unanimity among the 27, and we cannot rely forever on the sole action of the European Central Bank, whose democratic and parliamentary supervision must also be strengthened.

All these debates will to some extent take place, but they are rendered largely inaudible by the fragmentation of the candidacies on the left. The fact that the leaders concerned (*Insoumis*, Socialists, Ecologists, Communists, etc.) do not understand that what brings them together is far more important than what separates them is appalling. If we want to save the presidential election, it is urgent that the different candidates meet to debate what they have in common and their differences and that the left-wing voters arbitrate these differences between now and January.

The weakness of the current debate also shows once again the evils of French-style presidentialism. We will definitely not return to the indirect election of the president, and full proportional representation is not a panacea either. Beyond the necessary strengthening of the rights of parliament and the inversion of the electoral calendar, the French democratic system needs to be ventilated by introducing new forms of citizen participation, in particular with the popular initiative referendum. The number of signatories required under the 2008 constitutional revision is absurdly high, and only a new revision could unblock the situation. The presidential debate in 2022 could be an opportunity to move forward on this point. This is provided that the key issue of political campaign financing is also included, as it risks corrupting referendum democracy as much as representative democracy if not properly addressed. Proposals have been made to drastically reduce the weight of private donations and to introduce vouchers for democratic equality. They have started to be taken up by candidates and parliamentarians, and there is no reason why they cannot be adopted even before the next elections, which could help restore faith in politics.

The general lesson is clear: In order to save the presidential election, citizens and elected representatives of all tendencies must also and above all mobilize to overcome presidentialism.

The New Global Inequalities
December 14, 2021

What can we learn from the new *World Inequality Report 2022* published this week? The result of the contributions from over a hundred researchers from all continents, this Report, published every four years, allows us to examine the major fault lines in the world's inequalities. Beyond the now well-known findings on the rise of income inequalities over the last few decades, three main new features can be identified, relating to wealth, gender, and environmental inequalities.

Let us start with wealth. For the first time, thanks to the work of Luis Bauluz, Thomas Blanchet, and Clara Martinez-Toledano, researchers have gathered systematic data that allows for a comparison of wealth distributions in all countries of the world, from the bottom of the distribution to the top. The overall conclusion is that wealth hyper-concentration affects all world regions (and it has worsened during the Covid pandemic). At the global level, in 2020 the poorest 50% of the world's population owned just 2% of total private property (real estate, business and financial assets, net of debt), while the richest 10% own 76% of the total.

Latin America and the Middle East have the highest levels of inequality, followed by Russia and sub-Saharan Africa,

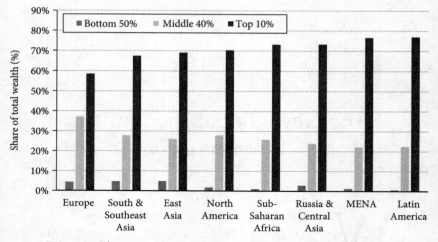

Figure 14. The extreme concentration of capital: wealth inequality across the world, 2021. The top 10% countries in Latin America capture 77% of total household wealth, compared with 1% captured by the bottom 50%. Net household wealth is equal to the sum of financial assets (e.g., equity or bonds) and nonfinancial assets (e.g., housing or land) owned by individuals, net of their debts.

Sources and series: wir2022.wid.world/methodology.

where the poorest 50% own just 1% of everything there is to own, while the richest 10% own around 80%. The situation is slightly less extreme in Europe, but there is really nothing to be proud of: The poorest 50% own 4% of the total, compared to 58% for the richest 10%.

Faced with this observation, several attitudes are possible. We can wait patiently for growth and market forces to spread the wealth. But given that more than two centuries after the Industrial Revolution the share held by the poorest 50% is barely 4% in Europe and 2% in the United States, we may be waiting a long time. It can also be argued that the current situation is the best we can do, and that any attempt to redistribute wealth would be economically dangerous. The argument is weak. In Europe, the share held by the richest 10% was 80–90% of total wealth until 1914. It has fallen in a century to less than 60% today, mainly to the benefit of the 40% of the population between the top 10% and the bottom 50%. This wealthy middle class was thus able to acquire housing and set up businesses, which greatly contributed to the prosperity of the Trente Glorieuses (or "Glorious Thirty," the period from 1945 to 1975 following World War II).

What can be done to prolong this long-term movement toward equality, which is historically inseparable from the evolution toward greater prosperity? Ideally, a redistribution of inheritance should be considered. At the very least, we need to stop promising tax giveaways to the wealthiest and focus on reforming the property tax, which is a very heavy and unfair tax for people on the way to home ownership, and which should become a progressive tax on net wealth.

The second lesson of the *World Inequality Report 2022* is gender inequality. Thanks to the data collected by Theresa Neef and Anne-Sophie Robillard, it is now possible to measure the evolution of women's share of total labor income for all countries in the world. This shows the extent to which gender inequalities remain high: At the global level, in 2020, women received barely 35% of labor income (compared to more than 65% for men). This share was 31% in 1990 and 33% in 2000: We can therefore see that progress exists but is extremely slow. In Europe, the share of women will reach 38% in 2020, which is still very far from parity.

This indicator gives a less watered-down and more accurate view of reality than the explanation for a given job: It shows precisely to what extent women do not have access to the same jobs and working hours as men, particularly as a result of multiple prejudices and discrimination and the lesser efforts made by the public authorities to structure the jobs in which women are most present (in particular in personal care, mass retailing, and cleaning jobs). The slow progress observed around the world over the last few decades also reflects the growing share of the wage bill captured by very high earners, who are overwhelmingly male. In some regions, such as China, there has even been a decline in the share of women in total labor income. All this calls for much more proactive measures than those adopted so far.

The third new feature of the 2022 Report is environmental inequalities. Too often, the climate debate is reduced to a comparison of average carbon emissions per country and their evolution over time. Thanks to the work of Lucas Chancel, we now have data on the distribution of emissions within countries and in different regions of the world. We can see that the poorest 50% of the world's population have relatively reasonable levels of emissions, for example 5 tonnes per capita in Europe. Meanwhile, the average emission reaches 29 tonnes for the top 10%, and 89 tonnes for the richest 1%. The conclusion is obvious: We will not meet the climate challenge by cutting everyone at the same rate. More than ever, the planet will have to take into account the multiple inequalities that cut across it to overcome the social and environmental challenges that undermine it.

Rightward Shift, Macron's Fault
January 11, 2022

How can we explain the rightward shift of the French political landscape? Even if the question is complex and admits of multiple answers, there is little doubt that the experience of Macronism in power bears an overwhelming responsibility.

Let us be clear: The dispersion of candidates on the left and the discouraging effect on voters also contribute to explaining this situation. However, this explanation is insufficient. If we take the total of all the left-wing candidates (Socialists, Ecologists, *Insoumis*, Communists, etc.), the total is a figure that is dismally low. According to the latest opinion poll carried out by *Le Monde* in December among 10,928 people, the total of those intending to vote for a left-wing candidate in the first round of the presidential election scheduled for 10 April 2022 is only just 27%, as compared with a total of 29% if we totalize the two far-right candidates (Zemmour and Le Pen), 17% for the right-wing candidate (Pécresse), and 24% for the incumbent president (who, for simplicity's sake, can be placed on the center-right, and is more and more perceived as such by the voters). Nowhere in our neighboring countries do we see such weakness of the left. Social democratic, socialist, labor, or democratic parties are in power in Germany and

Spain, or in a position to return to power in the next elections, in the United Kingdom and Italy.

It is true that the Parti socialiste has been in power in France for twenty of the last forty years, which may have given rise to a special kind of fatigue. By comparison, the Social Democrats were in power for only 7 years in Germany (1998–2005) and Labour for 13 years in the United Kingdom (1997–2010). Only in Spain have the Socialists been in power longer, which eventually fueled a split in the left with the emergence of Podemos, which both parties struggled to overcome, finally governing together. In France, the center-left should probably have recognized its mistakes in power and turned more to LFI (La France Insoumise) after the 2017 collapse. That might not have been enough, but there is still time to try.

The veer to the extreme right of the French political landscape can also be explained by other specific factors, starting with a particularly virulent post-colonial and Franco-Algerian trauma in France. Nostalgia for French Algeria and the breeding ground for xenophobia surrounding these wounds, which are still easily reopened, have thus played a central role in the emergence of both LePenism and Zemmourism.

All this is true, but it is not enough to account for the current situation. If France has become particularly right-wing, it is also and above all because Macronism in power has shifted a good part of the voters and elected officials from the center-left to the center-right, and even further and further to the right. On the economic front, Macron has applied the right-wing agenda: Abolition of the ISF (*impôt sur la fortune*, or wealth tax), a flat tax on dividends, deregulation of the labour market, absolute priority given to the "first in line" (*premiers de cordée*), with the consequences we know during the "yellow vests" (*gilets jaunes*) crisis, and a lasting discredit brought to any idea of a carbon tax in France. Having been robbed of its economic platform, the right then embarked on a chase with the far right, with considerable anti-migrant and anti-Muslim diatribes, as we witnessed during the LR primary.

The Macronist government itself, no longer knowing how to address the working classes, has begun to mimic the most extreme right. In particular, in recent years it has contributed to trivializing the nauseating rhetoric about the "Islamo-leftist gangrene in universities," a detestable phraseology that came from the far right before being taken up by a government that nevertheless relies in part on voters from the center-left. It has thus powerfully fueled the current right-wing trend, which it now wants to counteract, like an arsonist turned fire-fighter.

What can we conclude from all this? First of all, it would be healthy for Macron's supporters to realize this drift and draw the consequences. Either they approve of it and in that case vote for Pécresse: The difference between the two is infinitesimal, and this would restore clarity to the political landscape. It is all too easy for wealthy voters to have all the fiscal and financial advantages of Macronism while giving themselves a cheap conscience of so-called progressivism. After all, there is nothing nefarious about voting for a pro-business and slightly nationalistic right. Or they disapprove of this drift, and in that case they go back to vote for the left in the first round (there is no shortage of choice . . .).

Secondly and most importantly, all those who do not recognize themselves in this cynicism must come together to overcome their differences around a platform based on social, fiscal, and environmental justice. It is urgent to reorient the construction of Europe and the rules of globalization, and this will require a balance of power and unilateral measures (for example, on the minimum taxation of profits located in tax havens or on the carbon tax at the borders), but also constructive proposals of a social-federalist type, such as the creation of a European Assembly between the countries that wish to do so, with the power to vote for common taxes and to promote another development model. If the left abandons democratic and universalist internationalism and allows the market-based and the falsely European internationalism of the center-right (whether Macronian or Pécressian) to flourish, then it too will only contribute to preparing xenophobic nationalism to come to power in the more or less long term.

Sanction the Oligarchs, Not the People
February 15, 2022

The Ukrainian crisis has revived an old debate, namely, how to effectively sanction a state like Russia? Let's say it straightaway: It is time to imagine a new type of sanction focused on the oligarchs who have prospered thanks to the regime in question. This will require the establishment of an international financial register, which will not be to the liking of Western fortunes, whose interests are much more closely linked to those of the Russian and Chinese oligarchs than is sometimes claimed. However, it is at this price that Western countries will succeed in winning the political and moral battle against the autocracies and in demonstrating to world opinion that the resounding speeches on democracy and justice are not simply empty words.

Let us first recall that the freezing of assets held by Putin and his relatives is already part of the arsenal of sanctions that have been tried for several years. The problem is that the freezes applied so far remain largely symbolic. They only concern a few dozen people and can be circumvented by using nominees, especially as nothing has been done to systematically measure and cross-reference the real estate and financial portfolios held by each of them.

The United States and its allies are now considering disconnecting Russia from the Swift financial network, which would deprive Russian banks of access to the international system for financial transactions and money transfers. The problem is that such a measure is very poorly targeted. Just as with conventional trade sanctions, which after the 2014 crisis were largely instrumentalized by the government to strengthen its grip, the risk would be to impose considerable costs on ordinary Russian and Western businesses, with adverse consequences for the employees concerned. The measure would also affect a large number of bi-nationals and mixed couples, while sparing the wealthiest (who would use alternative financial intermediaries).

To bring the Russian state to heel, it is urgent to focus sanctions on the thin social layer of multi-millionaires on which the regime relies: A group that is much larger than a few dozen people, but much narrower than the Russian population in general. To give an idea, one could target the people who hold over 10 million euros in real estate and financial assets, or about 20,000 people, according to the latest available data. This represents 0.02% of the Russian adult population (currently 110 million). Setting the threshold at 5 million would hit 50,000 people; lowering it to 2 million would hit 100,000 (0.1% of the population).

It is likely that a considerable effect could already be achieved by targeting those with more than 10 million. These 20,000 people are those who have benefited most from the Putin regime since he came to power in 1999, and all the evidence suggests that a considerable proportion of their real estate and financial assets are located in Western countries (between half and three quarters). It would therefore be relatively easy for Western states to levy a heavy tax on these assets, say at a rate of 10% or 20% to start with, freezing the rest as a precaution. Threatened with ruin and a ban on visiting the West, let's bet that this group would be able to make itself heard by the Kremlin.

The same mechanism could have been used in the wake of the Chinese political crackdown in Hong Kong and could be applied in the future to the 200,000 or so Chinese holding more than 10 million euros. Although their assets are less internationalized than those of the Russians, they too would be hit hard and could destabilize the regime.

To implement this type of measure, it would be sufficient for Western countries to finally set up an international financial registry (also known as a "global financial registry" or GFR) that would keep track of who owns what in the various countries. As the *World Inequality Report 2018* has already shown, such a project is technically possible and requires the public authorities to take control of the private central depositories (Clearstream, Euroclear, Depository Trust Corporation, etc.) that currently register securities and their owners. This public register would also be an essential step in the fight against illicit flows, drug money, and international corruption.

So why has no progress still not been made in this direction? For one simple reason: Western wealthy people fear that such transparency will ultimately harm them. This is one of the main contradictions of our time. The confrontation between "democracies" and "autocracies" is overplayed, forgetting that Western countries share with Russia and China an unbridled hyper-capitalist ideology and a legal, fiscal, and political system that is increasingly favorable to large fortunes. In Europe and the United States, everything is done to distinguish useful and deserving Western "entrepreneurs" from harmful and parasitic Russian, Chinese, Indian, or African "oligarchs." But the truth is that they have much in common. In particular, the immense prosperity of multi-millionaires on all continents since 1980–1990 can be explained to a large extent by the same factors, and in particular by the favors and privileges granted to them. The free movement of capital without fiscal and collective compensation is an unsustainable system in the long term. It is by questioning this common doxa that we will be able to effectively sanction autocracies and promote another development model.

Confronting War, Rethinking Sanctions
March 15, 2022

So war is back in Europe, in its most brutal form. A country with 45 million inhabitants is being invaded by its neighbor with three times the population and eight times the weapons. Looking at it from a distance, one might be tempted to compare the situation to the border wars which opposed France and Germany three times between 1870 and 1945. Russia considers Crimea and the Donbass to be its property, as did Germany with Alsace and Moselle.

However, there are several key differences. The demographic and military imbalance is even more marked this time (Germany was 60% more populated than France in 1870, 1914, and 1940), and the authorities in Kiev have already indicated that they are ready to discuss the political status of the disputed territories, while respecting the rights of the populations concerned. In absolute terms, one could imagine a democratic and peaceful process, as much as possible on such sensitive issues. The problem is that the Russian state is using this border conflict as a pretext to invade and destroy the whole country and to challenge the very existence of the Ukrainian state. From this point of view, we are closer to the German invasion of the Second World War than to the confrontations of 1870–1871 or 1914–1918.

The Western response to this dramatic situation has so far been totally insufficient. In particular, European countries have the means to immediately stop Russian gas and oil supplies. A German academic study has just demonstrated that an immediate halt to imports would cost a maximum of between 2% and 3% of German GDP. These hydrocarbons should never have been burned and are now financing the destruction of Ukraine. It is time to leave them in the ground. If we don't act immediately and radically, we may well regret it bitterly.

On military aid, the United States and Poland had promised planes to Ukrainian pilots to defend themselves against Russian bombing, but then changed their minds. Overall, this is probably the first conflict in history where economically and militarily much more powerful countries (NATO countries collectively have ten times the GDP of Russia, and five times the airforce capacity) have announced in advance that they will not intervene, no matter how much human or material destruction there is on Ukrainian soil. In 1853, during the Crimean War, France and the United Kingdom had gone to defeat the Russian empire in order to contain its expansion to the south. The disproportion of forces between the West and Russia is even greater today, and the choice is to do nothing.

The explanation most often given is that the nuclear threat now renders the conventional weapons gap inoperative and prevents the use of conventional weapons. This argument is not entirely convincing and will require some explanation. If taken literally, it would imply that we would also have to stand idly by in the face of a similar invasion of other territories, no matter how destructive.

The most convincing explanation for this military hesitancy is that the European powers remain deeply traumatized by the cycle of nationalistic and genocidal self-destruction they experienced between 1914 and 1945, and have decided since 1945 to turn to the weapons of law, economics, and justice.

This is basically a positive development, up to a certain point, and on condition that these new weapons are fully used. This implies not only an immediate end to the financing of the Russian state through hydrocarbon purchases, but also a complete rethink of the functioning of economic sanctions, which today have a far greater impact on millions of ordinary Russians than on the small oligarchic and kleptocratic class on which the regime relies. It is claimed that the sanctions are aimed at the oligarchs, but the truth is that only a few hundred people are affected, without systematic control and with multiple loopholes, whereas tens of thousands of Russian fortunes invested in Western financial and real estate circuits should be targeted.

The stakes are high, not only to bring the Putin regime to its knees, but also to convince Russian and international opinion that the great speeches on justice and democracy are not empty words. In both Africa and Asia, more than half of the countries (and three quarters of the world's population and GDP by 2100) have abstained at the UN. Western countries are suspected of forgetting all their past invasions and thinking as always only of defending their interests and domination.

The problem is that the legal and financial system put in place by the West for several decades is primarily aimed at protecting the wealthy, wherever they come from, at the expense of others. If an ordinary Russian loses half of his pension or salary because of the fall in the ruble and inflation caused by the sanctions, then there is no recourse, no court where he can complain. On the other hand, if you want to deprive an oligarch with 100 million euros of half his fortune, then there are multiple procedures to challenge the decision, and very often you don't pay anything. We are so used to this that we don't pay attention to it anymore, but it is actually a totally biased and asymmetric rule of law. It is by going much further in law and justice that Western countries will be able to contribute to building a post-militaristic and post-colonial world.

The Difficult Return of the Left-Right Divide
April 12, 2022

In the first round of the 2017 presidential election, four candidates had achieved between 20% and 24% of the vote: This means that many second rounds were possible and could have occurred, within a deeply fragmented political and ideological landscape. Until the last moment, the voters of 2022 also had to face considerable uncertainties, and in particular a choice between a second round between the extreme right and the right (Le Pen against Macron, which the vast majority of voters now and quite logically place on the right) or between the right and the left (Macron against Mélenchon). This choice is anything but trivial and carries with it considerable consequences for the kind of public deliberation that will occupy the country for a fortnight (and perhaps longer): A debate centered on the hunt for immigrants and Muslims in the first case, or the hope of a discussion on wages and working conditions, health and education, social and fiscal justice, renewable energy and public services in the second.

However, whatever the outcome of the election, we can already be sure of one thing: We will not see the peaceful return of a reassuring left-right divide. Firstly, because the general movement to the right on the political landscape and the emergence of a powerful anti-migrant and anti-Muslim electoral bloc correspond to a strong trend, which Macronism in power has dangerously accentuated. Secondly, because it will take a long time for the forces of the left to unite and come to power.

Let's start with the first point. Things are now written. By appropriating the economic program of the right, Macron's centrism has not only become more right-wing, it has also contributed to making the country more right-wing, by pushing the Republican right into a dead-end chase with the far right on identity issues. The most dangerous thing is the arrogance of the president-candidate, who claims to be re-elected without any debate or program, or with botched measures that betray his fundamental elementary reaction to govern first and always for the first in line, by banking on the divisions of his opponents.

The ultimate in cynicism was reached with the issue of pensions. It should be remembered that to be entitled to a full pension in France, two conditions must be met: reaching the minimum legal age (currently 62) and validating the required length of contributions, which is increasing regularly and will soon reach 43 years (from the 1973 generation). In other words, for all those who have a higher education and start working at age 22 or older, raising the legal age to 65 years will have strictly no effect: Under the current legislation they will already have to wait until age 65 or older to have a full pension. On the other hand, for those who started working at 18, they will now have to wait until 65, or 47 years of contributions, even though their life expectancy is lower than that of the former. To propose such a reform, while claiming that long careers will be spared, even though they are by definition the only ones who will be affected, is a gross lie. By behaving in this way, Macron allows

Le Pen to present herself cheaply as the defender of the working classes and those who work hard.

The same thing happens when Le Pen proposes to reintroduce (in homeopathic doses) the tax on top financial wealth. The measure is largely hypocritical, since it also provides for the complete exemption of principal residences: Multimillionaires owning a château in Saint-Cloud will be entitled to a sharp drop in their real estate wealth tax, while ordinary French people suffer property tax increases. But as long as Macron refuses to re-tax high financial assets, this too allows Le Pen to present herself as a popular candidate at low cost.

This explosive political cocktail of violent anti-migrant rhetoric and social measures for the white working classes has already worked successfully in Poland and Hungary. Further afield, it is also what allowed the Democrats to regain power after the US Civil War, with a platform that was segregationist toward blacks but more social than the Republicans toward whites (including toward Irish and Italian migrants). The risk today is that such a social-differentialist (or social-racist) posture will prevail in France. In concrete terms, if Macron does not urgently make a strong social gesture, on pensions and tax justice, then his arrogance may cause him to lose a second round against Le Pen, or at least may lead her to reach very high and frightening scores.

Let's come to the second point. For the left to regain power, it will have to reconcile the working classes of different origins, which are today deeply divided, and therefore bring back to it those who no longer believe in social and economic promises and who rely on anti-migrant measures to change their lot. This will require an ambitious program for the redistribution of wealth and a sincere mea culpa on the errors of power. This will take time, because the rupture with the working classes is long-standing. The different parties

(*Insoumis,* Ecologists, Socialists, Communists, etc.) will have to overcome their resentments and come together again in a new popular, democratic, and internationalist federation. You cannot criticize presidentialism while at the same time refusing internal democracy when it comes to choosing your candidates. You cannot advocate internationalism while limiting your defense of democracy to national borders. All the more reason to start working on this now.

The Return of the Popular Front
May 10, 2022

Let's say it straightaway: The agreement reached by the French left-wing parties under the label of the "New Popular Union" is excellent news for French and European democracy. Those who see in it the triumph of radicalism and extremism have clearly understood nothing of the evolution of capitalism and the social and environmental challenges we have been facing for several decades. In reality, if we look at things calmly, the transformation program proposed in 2022 is rather less ambitious than those of 1936 or 1981. Rather than give in to the prevailing conservatism, it is better to take it for what it is: a good starting point on which to build further.

The program adopted marks the return of social and fiscal justice. At a time when inflation has already begun to cut into the incomes and savings of the most modest, it is urgent to change course. Those who claim that whatever-it-takes-type policies will not be paid by anyone are lying to the citizens. In order to compensate the most vulnerable for the effects of inflation and to finance investments in health, education, and the environment, it will be essential to ask the wealthiest to contribute.

Between 2010 and 2022, according to *Challenges* magazine (which cannot be suspected of leftism), the 500 largest French fortunes rose from 200 billion to nearly 1,000 billion, that is, from 10% of GDP to nearly 50% of GDP. The increase is even greater if we widen the focus and look at the 500,000 biggest fortunes (1% of the adult population), which now exceed 3,000 billion euros (6 million euros per person according to the World Inequality Database), compared with barely 500 billion for the poorest 25 million (50% of the adult population, each holding 20,000 euros on average). Choosing, in the midst of such a period of spectacular prosperity for the highest wealth and stagnation for the least prosperous, to abolish the meager wealth tax, when it should obviously have been increased, shows a curious sense of priorities. Historians who study this period will not be kind to the Macron governments and their supporters.

The first virtue of the left-wing parties is to have overcome their conflicts in order to unite in their opposition to this drift. In addition to reinstating the ISF (*impôt sur la fortune*), or wealth tax, it is proposed to transform the property tax into a progressive tax on net wealth, which would allow large tax reductions for millions of over-indebted French people from the working and middle classes. To encourage access to property, the package could eventually be completed by a minimum inheritance system for all.

The agreement reached between the *Insoumis* and the Socialists also provides for the extension of wage rights to gig workers and the strengthening of the presence of employees on boards of directors. Such a system has existed since the post-war period in Sweden and Germany (with up to 50% of the seats in large companies) and has allowed a better involvement of all in long-term investment strategies.

Unfortunately, it remains embryonic in France: The right has always been hyper-hostile to it (the Gaullists sometimes pretended to favor profit-sharing, in reality a few crumbs, but without ever questioning the monopoly of shareholder power), and the left has for a long time staked everything on nationalizations (as in 1981). The current shift toward a less state-centered and more participatory approach is reminiscent of the collective agreements of 1936 and opens the way to a new paradigm. Again, in the long run, much more should be done, for example by guaranteeing 50% of the seats for employees in all companies (small and large) and by capping the voting rights of an individual shareholder in large companies at 10%.

Let's turn to the European issue. All the "New Popular Union" parties advocate social and fiscal harmonization in Europe and a move to majority rule. To try to pass them off as anti-European, when they are the most federalist of all, is a crude maneuver. Liberals who claim to be European are in fact simply using the European idea to pursue their anti-social policies. In doing so, they are the ones putting Europe at risk.

If the working classes voted massively against Europe in the referenda of 1992 and 2005, and again in the Brexit vote in 2016, it is mainly because European integration as it has been conceived so far structurally favors the most powerful and mobile economic actors, to the detriment of the most fragile. It is Europe that has led the world and the United States in the chase for ever lower taxation of multinational profits, to the point where some people are now congratulating themselves on a minimum rate of 15%, barely higher than the Irish rate of 12.5%, with, moreover, multiple circumventions, and in any case much lower than what SMEs (small and medium-sized enterprises) and the middle and working classes pay. It is a lie to claim that the problem will be solved by remaining within the unanimity rule.

To put an end to fiscal, social, and environmental dumping in Europe, we must both make specific social-federalist proposals to our partners and take unilateral measures to break the deadlock. For example, as the European Tax

Observatory has shown, France could already impose a minimum tax rate of 25% or 30% on companies based in tax havens and selling goods and services in France.

Let us hope that the legislative campaign will be an opportunity to move beyond caricatures and make progress on these essential issues.

Moving Away from Three-Tier Democracy
June 14, 2022

Is it possible to break out of the present three-tier democracy in France and more generally on a European and international scale, and rebuild a left-right divide centered on questions of redistribution and social inequality? This is the central issue of the current legislative elections in France.

Let us first recall the contours of the three-tier democracy, as expressed in the first round of the presidential elections. If we add up the various candidates from the left-wing and ecological parties, we obtain 32% of the votes for the left-wing bloc, which can be described as being in favor of social-planning or social-ecological. If we combine the votes cast for Macron and Pécresse, we also obtain 32% of the votes for the liberal or center-right bloc. We arrive at exactly the same score of 32% if we add up the three candidates of the nationalist or extreme right bloc (Le Pen, Zemmour, Dupont-Aignan). If we divide the 3% of the unclassifiable ruralist candidate (Lassalle) between the three blocs, we arrive at three almost perfectly equal thirds.

This tripartition is partly explained by the specificities of the electoral system and the political history of the country, but its underpinnings are more general. It should be noted that the three-tier democracy does not mean the end of political divisions based on social class and divergent economic interests, quite the contrary. The liberal bloc achieves by far its best scores among the most socially advantaged voters, whatever the criterion used (income, wealth, education), especially among older people. The fact that this "bourgeois bloc" managed to attract a third of the vote is also due to the evolution of participation, which has become much higher among the wealthiest and oldest voters than among the rest of the population over the last few decades, which was not the case before. De facto, this bloc has combined the economic and wealthy elites, who used to vote for the center-right, with the educated elites who have taken over the center-left in many places since 1990, as the World Political Cleavages and Inequality Database shows.

With equal participation across all socio-demographic groups, however, this bloc would only garner about a quarter of the vote and could not claim to govern alone. In contrast, the left-wing bloc would be in the lead, as it scores best among the lower socio-economic classes, and especially among the younger generation. The nationalist bloc would also increase its score, but only slightly, as its lower-class vote profile is more evenly spread among the age groups.

In a way, one could say that this tripartition recalls the three great ideological families that have structured political life for more than two centuries: liberalism, nationalism, and socialism. Since the Industrial Revolution, liberalism has been based on the market and the social disembeddedness of the economy, and has mostly attracted the winners of the system. Nationalism responds to the resulting social crisis by reifying the nation and ethno-national solidarities, while socialism attempts, not without difficulty, to promote universalist emancipation through education, knowledge, and power sharing.

More generally, it has always been known that political conflict is structurally unstable and multidimensional (identity and religious cleavage, rural-urban cleavage, socio-economic cleavage, etc.) and cannot be reduced to an eternal one-dimensional left-right conflict reproducing itself identically over time. However, in many of the configurations observed in the past, or at least in those that have been retained, the social question took precedence and defined the main axis of the political conflict, opposing a social-internationalist left to a liberal-conservative right.

The novelty of the current situation is that the social question has lost its intensity, partly because the left in power has watered down its transformative ambition and has often rallied to the liberalism that has triumphed since the fall of communism, so that the question of identity has taken over.

What defines the three-tier democracy is first of all that the working classes are deeply divided around the migratory and post-colonial question: The young and urban working-class electorate has a more ethnically mixed sociability and votes for the left-wing bloc; conversely, the less young and more rural working-class electorate feels abandoned and turns to the nationalist bloc. The bourgeois bloc hopes to remain in power in perpetuity thanks to this division, but this is a risky and dangerous gamble, as the rhetoric deployed by the nationalist bloc (and often encouraged by the bourgeois bloc) leads to no constructive outcome and only exacerbates dead-end conflicts.

Contrary to what the other two blocs claim, the left-wing bloc is by no means unaware of the question of insecurity: On the contrary, it is the most capable of gathering the fiscal resources to strengthen the police and the justice system. As for the accusation of communitarianism, it is particularly inane. If young people of immigrant origin vote massively for the left-wing bloc, it is because it is the only one to defend them against the prevailing racism and to take the question of discrimination seriously.

The return to a confrontation centered on the social question is a necessity, not because the working classes would always be right when confronted with the bourgeois bloc (it is never simple to fix the right cursor on the scale of redistribution), but because conflicts resolved through mediation by social class offer more grist for the mill and allow democracy to function. Let us hope that these elections will contribute to this.

For an Autonomous and Alterglobalist Europe
July 12, 2022

Will Europe manage to redefine its place in the world geopolitical order? With Russia's invasion of the Ukraine and rising tensions with China, circumstances oblige it to do so, but hesitations are emerging.

Let's say it from the outset: We must maintain the link with the United States, but on the condition that we gain autonomy and get away from the egoism and arrogance that too often characterize the Atlantic and Western discourse toward the rest of the world. Europe has never been so rich. It has more than ever the means and the historical duty to promote another model of development and wealth sharing, more democratic, more egalitarian, and more sustainable. Otherwise, the new Western alliance will not convince anyone in its self-proclaimed crusade against autocracies and the reign of evil.

With the United States, Europe does indeed share a comparable experience of parliamentary democracy, electoral pluralism, and a certain form of rule of law, which is no mean feat. This may justify remaining in NATO, insofar as the alliance helps to defend this model. In this case, electoral pluralism is much more firmly established in Ukraine than in

Russia, and it is unacceptable to let a more powerful country invade its neighbor and destroy its state without reacting.

Discussion of borders should not be excluded a priori, but it should be done within the framework of the rule of law and on the basis of the dual principle of self-determination and the equitable and balanced development of the regions concerned (which may exclude the secession of the richest; this is not the case here). If NATO members stand for clear principles, then military support for the Ukrainians against invasion and destruction is justified, even more than at present.

It is also essential to explicitly recognize the limits of the Western democratic model and to work to overcome them. For example, we must fight for an international justice that would allow the Russian military and its leaders to be prosecuted for war crimes, while constantly reminding ourselves that the same rules should also apply to all countries, including of course the US military and its actions in Iraq and elsewhere. The principles of democracy and the rule of law must prevail everywhere and all the time.

Another example: The US Supreme Court has held for nearly two centuries that slavery and then racial discrimination were perfectly legal and constitutional, and has just ruled in its recent string of reactionary decisions that going out into the streets armed was also legal and constitutional. We must denounce the archaic institutions that abound in the United States and Europe and stop presenting ourselves to the world as a perfect and final version of the democratic model. For example, the ownership on both sides of the Atlantic of almost all the media by a few billionaires can hardly be considered the highest form of press freedom.

More generally, the disproportionate hold of private money on political life is symptomatic of a low-intensity democratic model, and helps explain the lack of relevance of programs and the record abstention of the poorest in elections.

The Western countries will be in a better position to spread democratic principles if they set higher standards for themselves. They will also be more credible if they stop making pacts with the most disreputable regimes as long as this allows them to make a few more dollars. If no real sanctions have been taken against the Russian oligarchs, or indeed against the petro-monarchic fortunes guilty of the worst abuses, it is to defend the interests of the Western financial and real estate circuits that shelter these fortunes, in Paris or on the Côte d'Azur as much as in London, Switzerland, or Luxembourg. It is also because this would require a transparency of assets that would risk turning against the Western fortunes. When the Chinese regime destroyed electoral pluralism in Hong Kong before our eyes in 2019, the only European reaction was to propose a new investment treaty in order to go even further in the free movement of capital, without control and without counterpart.

In general, the Western approach lacks a credible discourse on economic and social justice on a global scale. If India, South Africa, Senegal, or Brazil need resources to develop, who is going to stop them from doing business with Russia? If the West does not propose a new sharing of wealth, then it is China that will manage to unite the South. It is time to move away from the logic of unfulfilled promises (in particular those made at the Paris summit in 2015 to help poor countries adapt to global warming) and move toward a logic of rights.

In concrete terms, each country must be able to have access to a share of the revenue from the world's most prosperous economic actors (multinationals, billionaires, etc.), in proportion to its population. Firstly, because every human being should have an equal minimum right to health, education, and development. Secondly, because the prosperity of the rich countries would not exist without the poor countries: Western enrichment in the past and Chinese enrichment today have always been based on the international division of labor and the unbridled exploitation of the planet's natural and human resources. It is time to become aware of this historical legacy and to draw the consequences for the future.

A Queen with No Lord?
September 13, 2022

With the death of Elizabeth II, it is tempting to talk about the immutability of British institutions, in contrast to France and its many revolutions and constitutions. In reality, things are more complex, and the two countries are closer than they sometimes imagine, including when it comes to their political systems and institutions. The United Kingdom has seen its share of constitutional revolutions and upheavals, including the fall of the House of Lords, which has been without real power since the People's Budget crisis of 1909. Deprived of its Lords, who until then had been the backbone of the government and of executive and legislative power (most of the prime ministers had come from them), the British monarchy has been nothing more than a facade, governed entirely by its House of Commons, at least until the shock of the Brexit referendum in 2016.

Let's start at the beginning. The country had its first "French revolution" in 1530, with Henry VIII's dissolution of the monasteries. In the same way as in France after 1789, but more than two centuries before, the Church's land was sold to nobles and the bourgeoisie—those who could afford to buy them. This meant the state could be bailed out, while contributing to the creation of a new class of powerful and unified private owners, ready to embark unhindered on agrarian and then industrial capitalism.

After the beheading of Charles I in 1649 and then a brief period of republicanism, the Crown had no choice during the "Glorious Revolution" of 1688 but to submit to the power of Parliament, clearly dominated by the House of Lords. In the nineteenth century, societal and labor uprisings and the rise of universal suffrage reinforced the House of Commons's legitimacy. The conflict between the two Houses became inevitable and was played out in two stages.

In the 1880s, Lord Salisbury, leader of the Tories and the House of Lords, unwisely put forward the "referendum" theory: The Lords would have not only the moral and political right but also the duty, if they thought it good for the country, to veto legislation passed by the Commons—except in cases where legislation had been shown to the country before the vote. As a result, in 1894, the Lords vetoed Gladstone's (the leader of the Liberals) plans for new legislation on Ireland, a moderately popular reform that had not been explicitly put before voters. This allowed the Conservatives to win the 1895 election and return to power. But Salisbury's recklessness soon became apparent.

Back in power under Lloyd George, the Liberals had their famous People's Budget adopted by the Commons in 1909, with an explosive cocktail of measures: The creation of a progressive tax on total income (the supertax, which was added to the quasi-proportional taxes that had been imposed on different categories of income since 1842); an increase in inheritance tax on the largest inheritances; and an increase in property tax, particularly on large landed estates.

The whole package made it possible to finance a new series of social measures, in particular concerning workers' pensions, in an electoral context in which it was necessary to make promises to the working classes. The whole thing was perfectly calibrated to gain public support, while at the same time being an unacceptable provocation for the Lords, especially since Lloyd George did not miss an opportunity to publicly mock the idleness and uselessness of the aristocratic class. The Lords fell for it and vetoed the "People's Budget."

Lloyd George then chose to double down by passing a new law in the Commons, this time a constitutional one, whereby the Lords would no longer be able to amend finance laws and their power to block other legislation would be limited to one year. The Lords unsurprisingly vetoed this planned suicide, and a new election was called, which resulted in another Liberal victory.

Under the Salisbury Convention, and under pressure from the king to appoint a new batch of Lords if they reneged on their promise (a nuclear weapon rarely used in history but decisive in a crisis), the Lords were forced to pass the new Constitution Act in 1911. It was at this point that the House of Lords lost all real legislative power. Since 1911, popular opinion is expressed at the ballot box and in the House of Commons, and that is the law of the land in the United Kingdom, while the Lords have had only a purely advisory and largely ceremonial role.

In 1945, the working-class party won the election and set up the National Health Service. At the same time, the French senate also lost its veto power, after decades of blocking many essential social reforms (starting with women's right to vote, adopted by the Chamber of Deputies in 1919).

With Brexit and the arrival of Liz Truss into Downing Street, the two countries seem to be drifting apart again. But we can bet that protests and the social crises to come will continue to hold many surprises for us. France and the United Kingdom will continue to learn from each other and may one day meet again if the European Union finally succeeds in its social and democratic revolution. God save democracy!

Rethinking Federalism
October 11, 2022

In the face of the geopolitical and climate crisis, the question of sovereignty is on everyone's lips. Each country is seeking to regain control of its destiny, its supplies, and its production chains. People are even talking about European sovereignty, sovereignist federalism, or federal sovereignism. A contradiction in terms? Not necessarily, but on condition that we agree on the content. To avoid the pitfalls of nationalism and empty shells, it is essential to rethink the question of federalism, which must become a tool in the service of the best social, fiscal, and environmental objectives on offer, and no longer a means of reducing the power of the states and promoting a logic of dumping and generalized competition between territories.

Let us look back. In its early days, European integration drew some of its inspiration from the federal-conservative and ordo-liberal ideas of Hayek and the Freiburg school, with a form of rigid constitutionalization of the principles of free trade and competition. After the unlimited state power and devastating excesses of Fascism, Bolshevism, and Nazism, it was a matter of framing national sovereignty and promoting the reconstruction of Europe on the basis of economic exchange, without a new political majority suddenly putting an end to it. The memory of the 1930s, when electoral democracy had shown its limits, and when the breakdown of trade between countries had aggravated the crisis and plunged the world into an abyss, was still bitter.

We should not force the issue: European integration has always combined multiple inspirations, within unstable compromises, incomplete changes in direction, and a multiplicity of possible trajectories. The same will be true in the future: The European Union is not a finished product, far from it. Between 1950 and 1980, Europe also relied on pragmatic forms of industrial planning, directed credit, and control of capital flows. It was in this context that France, Germany, and their neighbors were able to build powerful welfare states, with considerable collective investment in education, health, housing, infrastructure, and social protection, thus demonstrating to the world that it is not only possible but essential to combine economic prosperity with the socialization of wealth, respect for individual rights, and a strong state capacity, under the control of voters and citizens.

Then the movement toward free trade and free movement of capital accelerated in the 1980s and 1990s, culminating in 1992 with the Maastricht Treaty. The French Socialists played a key role, exchanging with the German Christian Democrats the deregulation of capital flows for the single currency. The gamble was that it was worth it, in the sense that the pooling of monetary policy, with a European Central Bank (ECB) taking its decisions by majority vote and thus able to ignore a German veto, would in the long term allow new areas of shared sovereignty. The gamble has been partly fulfilled: Without the action of the ECB after the crises of 2008 and 2020, it is possible that the European countries would have been torn apart in sterile games of competitive devaluation and in collective impotence in the face of global markets. The problem is that not everything can be solved with money (as current inflation shows) and that at the same time we have gone too far in the sacralization of financial markets and competition, including in transport and energy, with the harmful consequences that we see today. All this in a context where Chinese and global competition has changed scale, so that unbridled free trade has greatly increased industrial relocation and the feeling of abandonment.

What could a federalism of the best social deal look like? The Manifesto for the Democratization of Europe offers some answers. Those countries that wish to do so could set up a European Assembly drawn from their national parliaments and competent to adopt a budget for investment in the future (environment, training, social cohesion) financed by common taxes on profits and on the highest incomes, assets, and carbon emissions. This would take nothing away from the sovereignty of each country, which, pending the adoption of these social-federal measures, could impose conditions on its partners to protect itself against unfair competition and social, fiscal, and environmental dumping.

The challenge is to develop a core group within the EU based on such principles, without destabilizing the whole. The task is surmountable, building on the Franco-German Parliamentary Assembly created in 2019 and giving it real powers, while opening it up to other countries. Such a core group would form the embryo of a future European Parliamentary Union (EPU), which could eventually bring together all 27 EU countries, or perhaps one day even the 43 countries meeting this week to launch the European Political Community (EPC).

The fact remains that what is urgently needed now is a hard core, not a new empty shell. In the face of the coming crises, in particular the new recovery plan that will have to be adopted at the European level and the painful fiscal measures that will have to be taken to deal with the debt and rising interest rates, it would be illusory to imagine that the EU will be able to cope with the situation with the unanimity rule. Only a few countries will be able to take the lead. All the more reason to start now.

Redistributing Wealth to Save the Planet
November 8, 2022

Let's say it straight out: It is impossible to seriously fight global warming without a profound redistribution of wealth, both within countries and internationally. Those who claim otherwise are lying to the world. And those who claim that redistribution is certainly desirable, sympathetic, and so on, but unfortunately technically or politically impossible, are lying just as much. They would be better off defending what they believe in (if they still believe in anything) rather than getting lost in conservative posturing.

Lula's victory over the agribusiness camp certainly gives some hope. But it should not obscure the fact that so many voters remain skeptical of the social-ecological left and prefer to rely on the nationalist, anti-migrant right, both in the South and in the North, as the elections in Sweden and Italy have shown. For one simple reason: Without a fundamental transformation of the economic system and the distribution of wealth, the social-ecological program risks turning against the middle and working classes. The good news (so to speak) is that wealth is so concentrated at the top that it is possible to improve the living conditions of the vast majority of the population while combatting climate change, provided that we give ourselves the means for an ambitious redistribution. In other words, everyone will naturally have to change their lifestyle profoundly, but the fact is that it is possible to compensate the working and middle classes for these changes, both financially and by giving access to goods and services that are less energy-consuming and more compatible with the survival of the planet (education, health, housing, transport, etc.). This requires a drastic reduction in the level of wealth and income of the richest, and this is the only way to build political majorities to save the planet.

The facts and figures are stubborn. The world's billionaires have continued their stratospheric rise since the 2008 crisis and during Covid and have reached unprecedented levels. As the *Global Inequality Report 2022* has shown, the richest 0.1% of the world's population now own some 80 trillion euros in financial and real estate assets, or more than 19% of the world's wealth (equivalent to one year of global GDP). The share of the world's wealth held by the richest 10% accounts for 77% of the total, compared to only 2% for the poorest 50%. In Europe, which the economic elites like to present as a haven of equality, the share of the richest 10% is 61% of total wealth, compared to 4% for the poorest 50%.

In France, the 500 richest people alone have increased between 2010 and 2022 from 200 billion to 1,000 billion, or from 10% of GDP to almost 50% of GDP (i.e., twice as much as the poorest 50%). According to the available data, the total income tax paid by these 500 wealthy individuals over this period was equivalent to less than 5% of this 800 billion enrichment. This is consistent with the tax returns of US billionaires revealed last year by ProPublica, which show an average tax rate in the same range. By instituting a one-off 50% tax on this enrichment, which would not be excessive at a time when small, hard-earned savings are paying an inflationary tax of 10% per year, the French government could raise 400 billion euros. One can imagine other formulas, but the fact is that the amounts are dizzying: Those who claim that there is nothing substantial to be recovered from this simply cannot count. For the record, the government just vetoed this week a decision by the National Assembly to increase investment in the thermal renovation of buildings (12 billion euros) and in the rail networks (3 billion), explaining that we could not afford such largesse. This begs the question: Does the government know how to count, or is it putting the interests of a small class ahead of those of the planet and the population, which is in dire need of renovated housing and trains that arrive on time?

Beyond this exceptional taxation of the 500 largest fortunes, it is obviously the entire tax system that needs to be reviewed, in France as in all countries of the world. During the twentieth century, progressive income tax was a huge historical success. The 80–90% tax rates applied to the highest incomes under Roosevelt and for half a century (81% on average from 1930 to 1980) coincided with the period of maximum prosperity, innovation and growth in the United States. For a simple reason: Prosperity depends first and foremost on education (and the United States was far ahead of the world at that time) and has no need for stratospheric inequality. In the twenty-first century, we need to extend this legacy to a progressive wealth tax, with rates of 80–90% on billionaires, and put the top 10% of wealth on the tax rolls. Above all, a substantial part of the revenue from the richest should be paid directly to the poorest countries, in proportion to their population and their exposure to climate change. The countries of the South can no longer wait each year for the North to deign to meet its commitments. It is time to think of the world in the making, or it will be a nightmare.

Rethinking Protectionism
December 13, 2022

Should we have boycotted the World Cup in Qatar? Probably not. Since we have always agreed to participate in sports competitions with regimes far removed from social and electoral democracy, starting with China (2008 Olympic Games) and Russia (2018 World Cup), the boycott of Qatar would have been interpreted as a new mark of the hypocrisy of Westerners, always ready to give lessons to a few small countries when it suits them, while continuing to do business with all those who bring them enough money.

Even if we choose not to boycott, this does not mean, however, that we should not do anything. On the contrary: We must act on the commercial lever, which is far more effective than the sporting lever. It is time for each country to redefine the conditions of trade with other territories, according to universal criteria of justice that apply equally to all. In the case of Qatar, the violations of fundamental rights have been proven, whether it be women's rights, the rights of sexual minorities, or social and trade union rights. Should customs duties of 10%, 30%, 50% be introduced? Should sanctions be concentrated on certain goods or on capital transfers, so that it is above all the wealthy and ruling classes that pay the price?

It is not up to me to decide here: It is up to democratic deliberation to do so and to place the cursor at the right level.

What is certain is that the argument that we will never be able to agree, and therefore we should do nothing and just apply absolute free trade to everyone, is incredibly hypocritical, nihilistic, and anti-democratic. Out of fear of democracy, we end up sacralizing free trade and the free movement of capital without limits, without even trying to subject these rules to any collective objective. When the Chinese regime destroyed electoral pluralism in Hong Kong before our eyes in 2019, the EU's only reaction was to propose a new liberalization of investment flows to Beijing.

The second reason for redefining the trade regime is obviously the environmental crisis. In 2022, we will continue to trade with China and the rest of the world without even trying to apply customs duties corresponding to the carbon emissions linked to the transport and production of these goods, in flagrant contradiction with the climate objectives. The same applies to fiscal and social dumping: If a country exports goods without respecting a common minimum base, then it is not only legitimate but essential to impose tariffs to restore the balance.

The third reason is linked to the fact that each country has the right to choose a productive specialization and to protect the sectors that it considers strategic. The best example today is that of batteries and electric cars. After having done the same for solar panels, China is massively subsidizing its companies to take control of the sector. The United States has followed suit. Only Europe is lagging behind, as in the case of the French bonus of 6,000 euros for the purchase of an electric vehicle, which applies regardless of where it is produced, whereas the US bonus of $7,500 is reserved for batteries and vehicles produced in the United States.

Faced with this predicted social and industrial collapse, what can a country like France do? The only solution is for each country to set its own conditions for further economic and trade integration, in terms of respect for fundamental rights, the fight against climate and tax dumping, and the protection of strategic sectors. These conditions must include tariffs and subsidies depending on the place of production.

Some people will be startled by this: If France unilaterally adopts such rules, is this not a clear violation of the European treaties signed in the past? The answer is more complex. In parallel to any unilateral action, ambitious proposals for collective measures are needed, including a new form of social federalism. Europe must be at the service of the social betterment: Countries that wish to do so must be able to adopt together the additional trade, social, and fiscal policies that seem appropriate to them, but this must not prevent each country from adopting its own measures.

With regard to unilateral protective measures, European law is more ambivalent than it appears. Article 3 of the Lisbon Treaty states that the EU's objectives are democracy, social progress, and environmental protection. How does destroying industrial jobs by importing all the equipment from China, without any consideration of the social damage and carbon emissions involved, serve these objectives? Some will argue that our prosperity depends on free trade, forgetting that it is not thanks to China that European purchasing power has increased tenfold over the last century (Chinese trade only accounts for a few percent of this increase at best). In any case, the debate should be political, not legal. The fact that past governments signed treaties constitutionalizing free trade, at a time when sovereignty was feared in Europe and current issues were ignored, cannot lead to the hands of future generations being tied indefinitely. More than ever, the law must be a tool for emancipation and not for the preservation of positions of power. It is by rethinking federalism and protectionism that the current crisis can be overcome.

President of the Rich, Season 2
January 10, 2023

In 2023, will Emmanuel Macron once again fall into the wrong era by illustrating himself as president of the rich? Unfortunately, this is what is in store with the pension reform. During his first term, he had already chosen to focus on the "first in line" and the abolition of the wealth tax. The result was a powerful feeling of injustice that led to the *gilets jaunes* (yellow vests) movement of working people fed up with the new taxes on fuel that they were required to pay while the richest received checks. In just a few months, the government has thus succeeded in permanently undermining the very idea of a carbon tax in France, which, to be accepted, would have to exempt the most modest and require proportionally much greater efforts from the most affluent.

In general, solving the climate challenge requires the construction of new collective standards of social and fiscal justice. The rise in energy prices, the exclusion of polluting vehicles, the switch to all-electricity, the accelerated renovation of housing, and so forth, will put increasing pressure on the working and middle classes. Not to mention that resources will also have to be found to invest more in health and training, which is the key to a productive and sustainable economy. In order to preserve social cohesion, it will be essential to tax the richest, and this will have to be done on the basis of objective and visible indicators such as income and wealth.

More than ever, we live in a world that needs justice and transparency. You only have to turn on the internet or open a magazine to know that millionaires and executives are doing wonderfully well. In France, the 500 largest fortunes have risen in ten years from 200 billion to 1,000 billion euros. Taxing this exceptional increase in wealth at 50% would be enough to bring in 400 billion. The size of the available tax pool would be even larger if we widened the focus to the richest 500,000 taxpayers (1% of the population) or the richest 10% or 20%. All these groups will have to be taxed in a graduated way, according to principles of justice that will have to be debated in the open, starting from the top. Everyone knows these realities and injustices today, at least as well as at the time of the Revolution and the privileges of the nobility. It makes no sense to pretend that there is nothing substantial to expect from this side. Repeating over and over again that the ISF (*impôt sur la fortune*), or wealth tax, brought in less than 5 billion euros is tantamount to taking citizens for fools: This low yield reflects the choice of successive governments to exempt billionaires and to rely on bogus self-declarations, and it is precisely this choice that must be called into question. It is by tackling the issue of justice that we will get out of the current crisis.

However, with the pension reform, the government is preparing to do just the opposite. The stated objective is to make 20 billion savings per year by 2030, in order to finance the government's other priorities. The problem is that this 20 billion will fall entirely on the shoulders of the most modest. Currently, to receive a full pension, you need two conditions: The legal age of 62 and the required length of contributions, which is 42 years for those born in 1961–1962 (and will gradually increase to 43 by the 1973 generation). Let's take a person born in 1961 who will therefore be 62 in 2023. If he or she has studied to the level of a master's degree and started working at 23, then he or she will already have to wait until 65 to reach 42 years of pensionable service. In other words, the reform consisting in pushing back the statutory age to 64 or 65 will by definition have no impact on these people. Of the 20 billion, the most highly educated will contribute exactly zero cents. By construction, these 20 billion will be taken entirely from the rest of the population, notably from workers and employees, who are also those with the lowest life expectancy and who already suffer from a profoundly unfair system, since it is their contributions that finance the pensions of executives with high life expectancy.

The government can try to disguise things: The reality is that it has invented a regressive tax that will be imposed exclusively on the least qualified. When Elizabeth Borne announces that no one will have to contribute 47 or 48 years of pensionable service, she is only admitting that some will contribute at 45 or 46 years, namely those who started working at 19 or 20 years of age and often work in difficult jobs. By definition, all the mitigation measures can only be financed by the less educated themselves. This reality is so obvious that the reform has united against it not only the left but also most of the right: The RN (Rassemblement National), of course, but also a growing part of the LR (Les Républicains).

As for the argument that we have no choice but to follow our neighbors, it is particularly weak. Firstly, because foreign systems actually combine multiple parameters and are more complex than is claimed. Secondly, the fact that no country has properly taken into account the abysmal social inequalities in retirement does not justify persisting in this way. Is the fact that gendered wage inequalities exist everywhere a justification for doing nothing about them? It is time for the pension system to focus on small and medium-sized pensions, with a public dependency service that allows everyone to finish their lives with dignity. The means exist for this. Let's hope that the deputies in the parliament and the social movement will be able to convince the government of this.

Emerging from the Pension Crisis through Justice and Universality

February 14, 2023

February 2023 may go down in history as the month when India became more populous than China, whose population is expected to be around 700 million by 2100 according to the United Nations, close to Europe. We could also focus on the earthquake that has just hit Turkey and Syria, in a region devastated by wars and oil interests, or on the consequences of global warming in Pakistan or the Sahel, or on the glaring inadequacies of sanctions against Russian oligarchs and support for Ukraine. Instead, what are we talking about in France? A profoundly unfair pension reform that is out of touch with reality, when we could do so much better to prepare for the future, such as debating an ambitious energy renovation plan, an investment program in training and research that is finally up to scratch, and so on.

In order to face up to major challenges such as aging, it is certainly inevitable that everyone should be asked to contribute. But it must be done fairly. There is only one way of trying to convince public opinion of the justice of a reform: It must be demonstrated that the effort required represents a higher proportion of income and wealth for the richest than for the poorest. If you reject any principle of this nature, then not only are you turning your back on more than a century of debate and political practice aimed at constructing collective standards of tax justice, but above all you are placing yourself in an extremely fragile situation in terms of defining what your own standard of justice is.

From this point of view, the documents presented by the government to defend its project are particularly poor. All we know is that the increase in the age and duration of contributions will bring in 17.7 billion euros per year by 2030, without any breakdown by income level or by social class or profession. And with good reason: If the government were to present these figures, it would immediately become apparent that the richest are being asked to contribute at a much lower rate than the middle classes and the poorest. The reason is simple. Raising the statutory retirement age to 64 has by definition no impact on the most highly educated and senior managers: If you started working at 22 or 23, you already have to contribute 42 years (soon to be 43) and therefore have to wait until 64 or 65 to get a full pension. The acceleration of the transition to 43 years of contributions will certainly affect part of this group (only the over-50s), but much less than workers and employees who started working at 19 or 20: The latter will also bear the brunt of the postponement of the legal age and will need to have 44 years of contributions for a full pension (and sometimes 45 or more, whatever the government says), even though they are the ones who have the lowest life expectancy and finance the retirement of executives.

How to get out of this crisis? Three principles are essential: universality, progressiveness, and justice. The government no longer has a choice: It must rebuild the system on the basis of the same number of years of service for all. If it chooses 43 years of pensionable service, then this must apply to everyone, without exception. But beware: If the government is sincere in its approach, then by definition the legal age of 64 is no longer relevant. If you have 43 years of service, then you can take your full pension, full stop. The problem is that the government spends its time confusing the debate by pretending that it is going to improve the long career system, while introducing clauses so complex on the dates of validated quarters at 19 or 20 that they do not apply to anyone. The most enormous manipulation is the following. As a general rule, the annuities include two years per child (one of which can be attributed to fathers for children born since 2010), as well as an additional year in case of parental leave. However, these years for children are only very partially taken into account in the complex rules linked to the long career system. This is why the government is so keen to maintain the legal age barrier at 64: This is what allows it to demand de facto 44 or 45 years of service (or more) from working-class women who started working early, while women in senior management will be entitled to 43 years of service without difficulty. This game must stop: If the 43-year rule is announced, then it must be applied to everyone equally, without exception, that is, by eliminating the 64 years. Some people on the left or the right prefer 42 or 41 years of service. The debate is legitimate, but in any case the rule must apply to all.

The second principle is progressiveness. The government wants to raise the minimum pension to 85% of the net SMIC (minimum wage; 1,200 euros), but again with very restrictive conditions. It is time to apply explicitly progressive replacement rates, for example 100% at the SMIC level, 75% at 3 SMIC, and 50% at 6 SMIC. Discounts (*décotes*) should no longer be applied to the lowest pensions.

The third principle is fairness in financing. To achieve this, the CSG (generalized social contribution) must be extended. The CSG has had a progressive component since its creation in 1990, with a reduced rate for small pensions. Additional rates could be created on incomes above 5,000 or 10,000 euros per month, as well as a CSG rate of 2% on the 500 largest fortunes, which alone would bring in 20 billion euros per year, which is much needed for hospitals and pensions. One thing is certain: It is by renewing the spirit of justice that we will get out of the current crisis.

Macron, the Social and Economic Mess
March 14, 2023

Let's say it straight away: Macron is in the wrong era and is wasting our time. He is applying recipes that are completely unsuited to the world of the 2020s, as if he had remained intellectually stuck in the era of the market euphoria of the 1990s and early 2000s, the world before the 2008 crisis, Covid and Ukraine. Yet the current context is one of rising inequality, hyper-prosperity of wealth, and the climate and energy crisis. The urgent need is for investment in education and health and the establishment of a fairer economic system, in France and in Europe, and even more so on an international scale. But the government continues to pursue an anti-social policy from another age.

On pensions, Macron had tried in 2019 to promote the idea of a "universal" pension, with a unification of the rules between the schemes, which are indeed too complex. The problem is that he was supporting a very unequal universal pension, which roughly speaking perpetuates the abysmal inequalities of working life until death. Many other universal pensions are possible, with an emphasis on small and medium pensions, with a replacement rate varying with the level of salary, all financed by a progressive levy on income and wealth (with, for example, the introduction of a 2% CSG [generalized social contribution] rate on the 500 largest fortunes, which alone would bring in 20 billion euros).

Today Macron is no longer even trying to pretend and play the modernizer of the social state: The 2023 pension reform simply aims to raise money, without any objective of universality or simplification. It is even the most opaque of the parametric reforms that one could have imagined. The new rules on long careers are totally muddled. The so-called measure on small pensions at 1,200 euros will in the end concern less than 3% of pensioners, and it will have taken the government a year to arrive at this still very approximate figure, even though it has the entire state apparatus at its disposal and spends billions on consultancy firms. The reality, which it is now impossible to hide, is that the efforts will fall mainly on low- and middle-paid women, who will have to work two years longer in difficult and poorly paid jobs, when they are still in employment.

Beyond these injustices and all the time wasted on pensions, the social and economic mess of the Macron presidency is found in other areas. If we look at the evolution of higher education resources, we see that the budget per student has decreased by 15% in France over the last ten years. Rather than rehashing McKinsey PowerPoints on the start-up nation, the government would be well advised to meditate on the basic lesson of all economic history, namely that it is investment in training that is the source of prosperity. In general, the construction of the welfare state was a huge historical success in the twentieth century, and this achievement must be built upon.

It is thanks to a powerful movement of investment in education, health, and public infrastructure that we have achieved both greater equality and prosperity than ever before in history. Public resources mobilized in education have increased tenfold, from 0.5% of national income in Western countries before 1914 to around 5–6% since the 1980s–1990s.

In the middle of the twentieth century, the United States was by far the world's educational leader (with 80% of an age group in long secondary education by 1950, compared with 20–30% in France or the United Kingdom at the same time), and this is why it was also the economic leader. All of this was done with inequalities strongly compressed, thanks to tax progressivity: The top income tax rate reached 81% on average across the Atlantic from 1930 to 1980. Clearly, this has not harmed the exceptional productivity of the world's leading economy, quite the contrary.

The great lesson of history is that prosperity comes from equality and education, not from chasing inequality. Reasonable income disparities can be justified (say, one to five), but stratospheric inequalities serve no public good. This lesson has been forgotten, and social and educational investment has stagnated for 30 years, while student numbers have increased. We need look no further for the reasons for the stagnation of productivity.

By weakening the social state instead of expanding it, the government is weakening the country and its place in the world. It also misses a historical turning point, which is the transition from the social-national state to the social-global (or social-federal) state. In the twentieth century, the social state developed primarily within the national framework, sometimes superbly forgetting North-South inequalities. This is all the more problematic because Western enrichment could never have taken place without a very strong international integration and without the often brutal exploitation of the natural and human resources available on a global scale. It is no longer possible to ignore the consequences of the environmental damage caused by the enrichment of the North (including, of course, Russia and China).

The social-global state must be based on an overhaul of the global economic and fiscal system, with the richest global players (multinationals, billionaires) being taxed for the benefit of all. This is the way to revive the social state in the North as well as in the South and to get out of the current contradictions.

Can We Trust Constitutional Judges?
April 11, 2023

As the "wise men" of the Constitutional Council prepare to give their decision on pensions, it is worth asking a simple question: In general, can we trust constitutional judges? Let us be clear: Constitutional courts play an absolutely indispensable role in all countries. Unfortunately, like all powers, these precious and fragile institutions are sometimes instrumentalized and damaged by the people to whom these eminent functions have been entrusted, who often try to impose their own political preferences under the guise of law.

There are many examples in history. In the United States, the Supreme Court ruled in 1896 in the ominous *Plessy v. Ferguson* decision that it was perfectly legal for southern states to segregate as much as they wanted. The ruling formed the legal basis for the segregationist order until the 1960s. In the 1930s, the Court repeatedly censured social legislation passed by Congress under the New Deal, on the grounds that some of it constituted an unacceptable infringement of the freedom of enterprise (which the judges chose to interpret as they saw fit). Re-elected in 1936 with 61% of the vote, Roosevelt announced his intention to appoint new judges (the Constitution did not specify their number) in order to break the deadlock. The Court finally decided to give in and validate a decisive law on the minimum wage that it had previously censured. Closer to home, the *Citizens United* decision in 2010 and the *McCutcheon* decision in 2014 ruled that it was illegal—as contrary to the principles of "free speech"—to impose limits on private political funding.

In Europe, too, there is no shortage of abuse of power. A particularly extreme case is the Kirchhof affair in Germany. During the 2005 campaign, Paul Kirchhof, a tax lawyer who was very angry about taxes, was presented as Angela Merkel's future finance minister, with a shock proposal: A "flat tax" limiting the tax rate on the highest incomes. In the political sphere, everyone is of course free to express his or her own opinions, which in this case did not appeal to the Germans: All the indications are that this proposal contributed to reducing the CDU's score, so much so that Merkel was forced to form a coalition with the SPD and to separate from her adviser.

The interesting point is that in 1995, while he was a member of the Constitutional Court in Karlsruhe, the same Kirchhof had issued a ruling that any income tax above 50% was unconstitutional. The case caused a scandal, and the ruling was finally overturned in 1999 by the German constitutional judges, who confirmed in 2006 that it was not within their remit to set quantitative limits on tax rates.

In France, the Constitutional Council decided to censure a timid law on parity in 1982. It was not until 1999 that a constitutional amendment allowed the judges' decision to be overruled. More recently, Jean-Louis Debré explained that a tax rate of 75% above one million euros was unconstitutional, on the grounds that tax must remain a "contribution" and cannot become a "spoliation." The problem is that the constitution nowhere sets such a numerical limit, which is a matter of pure personal interpretation. Like any citizen, the former president of the Constitutional Council is of course free to consider that the 80%–90% rates applied in the United States from 1930 to 1980 did not produce the desired results (in this case they worked very well and in no way threatened the rule of law), or more generally that this is not a good policy in his eyes. But the fact that he can use his functions as a judge to make his point of view prevail, without even having to provide any serious argument, is a clear abuse of power.

Even worse: French constitutional judges have developed in the last decades a strange doctrine without any legal basis according to which wealth tax cannot exceed a certain percentage of income, even in the case where income represents a ridiculous percentage of the highest fortunes, which de facto prohibits any significant levy on wealth. Yet there is nothing in the constitution that prohibits taxes on property, even in the absence of any income, as evidenced by the inheritance tax and the property tax, which have played an essential role in the fiscal and social system since the Revolution. By acting as the objective accomplices of the owners, judges damage their function and democracy.

What can we conclude from all this? Firstly, we should not expect too much from constitutional judges. Like all human institutions, this power must be constantly questioned, evaluated, and controlled democratically, if necessary by constitutional, amendments. It should never be sacralized. With regard to the present decision, it is clear that the use for pension reform of the privileges reserved for social security financing laws is contrary to the spirit of the constitution and would merit a complete censure (which should have taken place earlier). The refusal of the referendum process by the judges would also open a serious crisis and confirm the muddled and unworkable nature of the 2008 constitutional revision. In any case, we must return to the essentials: Above all, it is democratic and social mobilization that produces historical change and allows the law to become a tool for emancipation and not for the preservation of positions of power.

What If Economists Were about to Change?
May 9, 2023

Let's celebrate. The American Economic Association (AEA), the main professional organization for economists in the United States, has just awarded the Clark Medal to Gabriel Zucman for his work on the concentration of wealth and tax evasion. Awarded each year to a winner under the age of 40, the distinction is given in particular for innovative work demonstrating the considerable importance of tax evasion by the richest, including in Scandinavian countries, which are often considered models of virtue. Endowed with an immense capacity for work, a rare attention to detail, and an unparalleled talent for unearthing new data and making it speak for itself, Gabriel Zucman has also revealed the unsuspected extent of corporate tax evasion by multinationals in all countries.

Now director of the European Tax Observatory, he devotes the same energy to finding solutions to the ills he documents. In one of its first reports, the Observatory demonstrated that EU Member States could choose to go further than the 15% minimum rate set by the OECD (which is too low and largely circumventable), without waiting for unanimity. By imposing a 25% rate on the profits of each multinational company wishing to export goods and services—the same as that paid by domestic producers—then France would gain an additional 26 billion euros in revenue and encourage other countries to do the same.

The fact that the AEA has chosen to reward this work is important because it shows that the core of the profession is beginning to realize that the current social and fiscal model is unsustainable. Let us not overstate the case: Economists have always been less monolithic than is sometimes imagined, including in the United States. In 1919, AEA president Irving Fisher chose to devote his presidential address to the issue of inequality. He told his colleagues that the growing concentration of wealth was becoming America's main economic problem, and if left unchecked, America would become as unequal as old Europe (then perceived as oligarchic and contrary to the American spirit). Fisher was distressed by Willford King's 1915 estimates that "2% of the population owns more than 50% of the wealth," and that "two-thirds of the population owns almost nothing," which he saw as "an undemocratic distribution of wealth" threatening the very foundations of American society.

It is in this context that the United States applied rates of over 70% at the top of the income hierarchy from 1918–1920 (under the Democrat Wilson), before any other country. When Roosevelt was elected in 1932, the intellectual groundwork had long been laid for the introduction of a high degree of tax progressivity, with the famous Victory Tax of 88% in 1942 and 94% in 1944. The United States would apply similar rates in Germany and Japan: In the spirit of the times, these fiscal institutions were seen as an indispensable complement to democratic institutions, without which the latter might sink into a plutocratic drift.

Unfortunately, these lessons have been forgotten, and the United States and much of the world has entered a new oligarchic spiral since the 1980s and 1990s. It would certainly be an exaggeration to blame only economists for this. If the counteroffensive launched in the 1960s and 1970s by Friedman and Hayek was able to bear fruit, it was also due to a lack of collective appropriation of the New Deal institutions among citizens and within the social and trade union movement. The intellectual battle was also played out in the philosophy departments: When Rawls published his "Theory of Justice" in 1971, he laid the conceptual foundations of an ambitious egalitarian program but remained relatively abstract in its practical implications. At the same time, Friedman and Hayek were perfectly clear on their objective of demolishing tax progressivity.

The fact remains that economists bear a particular responsibility for the deregulation and liberalization movement of recent decades. There are, of course, the effects of the search for private funding, which tends to right-wing the discourse. In 2016, when Sanders and Warren took up bold wealth tax proposals (with rates of up to 6–8% per year above $1 billion), former Clinton Treasury Secretary and Harvard President Larry Summers—a great advocate of absolute liberalization of capital flows—nearly choked and did not hesitate to violently attack researchers like Zucman who supported these proposals (which are, however, common sense, given the almost zero income tax rates paid by billionaires).

There are also intellectual reasons linked to the evolution of the discipline of economics. In order to give itself an autonomous scientific appearance, economics has tended to cut itself off from history and sociology and to naturalize the institutions studied (the market, property, competition), forgetting in the process their social and political embedding within particular societies. Mathematical models can be useful if they are used wisely and not as an end in themselves. Statistical techniques can be useful, provided that we do not lose sight of the critical view of sources and categories. There is still a long way to go before political and historical economy regains its rightful place in the social sciences.

For a European Parliamentary Union (EPU)

June 13, 2023

Faced with new social, climatic, and geopolitical challenges, Europe has no choice but to reinvent itself if it wants to play a useful role for its citizens and the planet. It is with this in mind that a new organization created in 2022 met a fortnight ago in Moldova: the European Political Community (EPC). The initiative deserves to be applauded. By bringing together 47 countries, from the United Kingdom to Ukraine and from Norway to Switzerland and Serbia, the EPC is a reminder that the 27-nation European Union (EU) is not set in stone forever. Increasingly advanced discussions and cooperation must extend to the whole continent and beyond, if only to assert and defend a minimum foundation of common political rules and principles, which is no mean feat. However, it is clear that the EPC comprises such a broad spectrum of countries that it will have even more difficulty than the EU in taking decisions and pooling the resources needed to move forward and influence world developments.

This is why it is essential to supplement this multi-circle architecture with a hard core made up of a small number of countries that are genuinely ready to go further in the process of political union. For the sake of clarity, this core group could be called the European Parliamentary Union (EPU). The EPU could be based on the Franco-German Parliamentary Assembly (FGPA) set up in 2019 when the bilateral treaty linking the two countries was renewed, but opening it up to all countries wishing to join and giving it real powers—whereas this Assembly currently plays a purely consultative role, which can also be seen as the start of an initial running-in period.

Ideally, the EPU should include at least France, Germany, Italy, and Spain, which together account for more than 70% of the population and gross domestic product of the eurozone. If there is no other solution, the EPU could also start with two or three countries. Ultimately, of course, the aim is to convince all 27 EU countries—or even the 47 members of the EPC and beyond—to join this core group. But this could take many years, essential if the EPU is to prove itself and demonstrate to Europe and the world that it is possible in the twenty-first century to conceive of a new form of social and federal union, transnational and democratic.

What powers and objectives would the EPU have? Generally speaking, the idea is to be based on the principle of the best social, environmental, and fiscal deal. In other words, the EPU must enable its members to go further in the field of social and environmental progress and fiscal justice, if a majority is in favor, without preventing its members from moving in this direction with their own resources. The EPU must deepen what has been Europe's great historic success since the post-war period, namely the consolidation of parliamentary democracy and the construction of the welfare state.

In concrete terms, the EPU would have the power to adopt a budget for investment in the future, with particular emphasis on energy and transport infrastructure, thermal renovation of buildings, and a massive investment plan in health, training, and research. This budget would be voted on by a European Assembly (EA) made up of parliamentarians from the various member countries of the EPU, in proportion to the populations and political groups present in the various national parliaments. The European Assembly would also have the power to borrow jointly to finance such a budget. Limits would naturally have to be set to such a power, and these would have to be specified in the intergovernmental treaty establishing the EPU, as well as the conditions for its revision. But it is essential to be able to take decisions more flexibly and reactively than is possible under the unanimity rules of the 27-member EU if we are to be able to deal with the many crises and challenges that lie ahead. In 2020, it took months to convince the 27 to launch a joint loan for the first time. It was only the exceptional urgency caused by the Covid crisis and the prolonged confinement of hundreds of millions of Europeans that broke the deadlock. This is not a calm or effective way of operating, and above all it will not enable us to meet the challenges of the future.

In line with the Manifesto for the Democratization of Europe launched in 2018, which collected more than 100,000 signatures, the EPU could also adopt common taxes on the profits of multinationals, the highest incomes and wealth and the highest carbon emissions.

More generally, the context of the last few years makes it possible to go even further and replace the defunct "Maastricht consensus" with a new social and democratic vision of the European project, in line with the workers' mobilizations of the last two centuries. For example, it would be conceivable for member states to entrust the EPU with the task of laying down minimum rules for employee representation in corporate governance or the system for financing the media and election campaigns.

It would be absurd to pretend that such a path is well trodden: Almost everything remains to be invented. However, two points need to be emphasized. Firstly, the alternative option of imagining a rapid overhaul of the 27-member European treaties is wishful thinking. Secondly, nothing in the current rules prevents a hard core of European countries from moving forward. All the more reason to get on with it now.

France and Its Territorial Divides
July 11, 2023

To analyze the urban riots of 2023—by far the most serious since those of 2005—and the political misunderstandings to which they give rise, it is essential to go back to the roots of France's territorial malaise. The suburbs that are currently catching fire have much more in common with the abandoned villages and midsize towns than is sometimes imagined. The only way out of the current contradictions is to bring these different disadvantaged areas together politically.

Let's look back. Between 1900–1910 and 1980–1990, territorial inequalities decreased in France, both in terms of differences in gross domestic product per capita between departments and inequalities in property wealth or average income between communes and departments. The opposite has been true since 1980–1990.[1] The ratio between the GDP per capita of the five richest and poorest *départements,* which had fallen from 3.5 in 1900 to 2.5 in 1985, has risen to 3.4 in

---

1 J. Cagé and T. Piketty, *A History of Political Conflict: Elections and Social Inequalities in France, 1789–2022* (Belknap Press of Harvard University Press, 2025).

2022. We are witnessing an unprecedented concentration of GDP in just a few *départements* in the Île-de-France region (notably Paris and Hauts-de-Seine), linked to the unprecedented expansion of the financial sector and the headquarters of major companies, and to the detriment of provincial industrial centers. This spectacular development has been exacerbated by financial deregulation and trade liberalization, as well as by public investment benefiting the capital region and major cities (TGV versus regional trains).

Similar trends can be seen in the inequalities between municipalities. The ratio between the average property wealth of the richest and poorest 1% of municipalities has risen from 10 in 1985 to 16 in 2022. In Vierzon, Aubusson, and Château-Chinon, the average property value is barely 60,000 euros. It exceeds 1.2 million euros in the 7th arrondissement of Paris, as well as in Marne-la-Coquette, Saint-Jean-Cap-Ferrat, and Saint-Marc-Jaumegarde. The ratio between the average income of the richest and poorest 1% of municipalities has risen from 5 in 1990 to over 8 in 2022. The average income in Creil, Grigny, Grande-Synthe, and Roubaix is barely 8,000–9,000 euros per inhabitant per year. It reaches 70,000–80,000 euros in Neuilly-sur-Seine, Le Vésinet, or Le Touquet. It even exceeds 100,000 euros per inhabitant (including children!) in the 7th and 8th arrondissements of the capital.

The central point is that there are considerable inequalities between communes across the country, both within large conurbations and between midsize towns and villages. At the top of the territorial hierarchy are the wealthiest suburbs of the major metropolises, some of the city centers, and a number of wealthy towns and villages. At the very bottom of the pyramid, the poorest suburbs have been hard hit by deindustrialization. They are now just as poor as the poorest towns and villages, which was not the case historically. These different disadvantaged areas certainly face specific challenges. Poor suburbs have much more experience of diversity of origin and proven discrimination in police practices and access to housing and employment. There is an urgent need for public authorities to finally acquire the means to objectivize and rigorously measure changes in this discrimination—the existence of which has been demonstrated by a multitude of research studies.

The various disadvantaged areas are also characterized by their specific integration into the productive structure. The poorer suburbs include a large number of service workers (retail, catering, cleaning, health, etc.) who continue to vote for the Left. Conversely, the poorer towns and villages now include more blue-collar workers exposed to international competition. Many of them have felt abandoned by the left- and right-wing governments of recent decades (accused of having staked everything on European and global trade integration, with no limits and no rules) and have joined the National Front–National Rally. But contrary to what the political leaders of the nationalist bloc imagine, these voters expect above all socio-economic answers to their problems, and not a strategy of identity-based confrontation, which in no way corresponds to the real state of French society, as shown by the very high levels of mixing and intermarriage.

The truth is that poor suburbs and poor towns and villages have much in common with the richest areas, particularly in terms of access to public services and municipal budgets. The reason is simple: The resources available to local authorities depend first and foremost on local tax bases, and the national measures supposedly put in place to tackle these abysmal inequalities have only ever reduced a small part of them. In the end, the per capita budget is higher in richer municipalities than in poorer ones, so that public money exacerbates the initial inequalities instead of correcting them, in all good conscience. The proposals made in 2018 by the Borloo report to objectify this reality and put an end to it have been abandoned, and the liberal bloc continues to explain today that no further redistribution is conceivable. Faced with the impasses of the other two blocs, it is now up to the left-wing bloc to rally the disadvantaged territories around a common platform.

Who Has the Most Popular Vote or the Most Bourgeois Vote?
September 19, 2023

The question of the popular or bourgeois profile of different votes has always given rise to a great deal of controversy. In *A History of Political Conflict*, Julia Cagé and I develop a method for establishing a number of facts and trends. We begin by compiling the electoral results at muncipality level for all legislative and presidential elections from 1848 to 2022, as well as for the most significant referendums from 1793 to 2005. We then classify the 36,000 municipalities (*communes*) according to their average wealth, from the poorest 1% of *communes* to the richest 1%, and observe how the score obtained by the various candidates and political currents evolve in proportion to their national average score. We use several wealth indicators, in particular the average income per commune. We obtain the same results with other indicators such as the average value of housing.[1]

---

1 J. Cagé and T. Piketty, *A History of Political Conflict: Elections and Social Inequalities in France, 1789–2022* (Belknap Press of Harvard University Press, 2025).

When it comes to the Macron or Ensemble vote in recent elections, we see an exceptionally steep slope. We sometimes find right-wing votes that are even more bourgeois than Macron, for example Madelin in 2002 or Zemmour in 2022 (proof if it were needed that the anti-immigrant vote is by no means the exclusive preserve of the working classes), but these are smaller votes in terms of electorate size. For votes of comparable importance (let's say around 20–30% of the vote or more in the first round), so if we compare it to the Giscard, Chirac, Balladur, De Gaulle, or RPR-UDF votes in the past, then the Macron or Ensemble vote appears to be more bourgeois, in the sense that it has approximately the same slope as past right-wing votes at the top of the distribution (within the richest communes), but a steeper slope at the bottom of the distribution (within the poorest communes). In other words, while the traditional right managed to capture part of the vote in the most modest *communes*/local authorities, particularly in rural areas, this is not the case for the Macron vote.

It should be pointed out that this trend began before Macron. For example, the Sarkozy vote in 2007 or 2012 is steeper than the Giscard or Chirac votes of the past, particularly at the lower end of the distribution, because part of the rural working-class electorate voting for the right has already begun its transition to the National Front–National Rally (FN-RN), particularly as a result of the disappointment following the 2005 referendum and the parliamentary ratification of the Treaty of Amsterdam. Which goes to show that the FN-RN vote in small towns and villages is above all a socio-economic vote, concerned about de-industrialization and international trade integration, and not an identity vote that can be captured with facile rhetoric about the *karcher* or *racaille*. Basically, Macron is simply extending and amplifying this bourgeois-Sarkozist evolution.

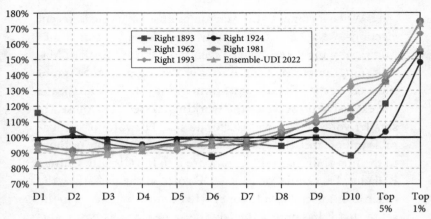

Distribution of the population by deciles as a function of municipal income per inhabitant

Figure 15. Is the Ensemble vote the most bourgeois in French electoral history? The vote for the Ensemble–L'Union des démocrates et indépendants (UDI) bloc in 2022 increases strongly with income. The slope is on the whole comparable with the vote profiles for the right wing observed in the past, with the difference that the latter generally got better scores in the poorest municipalities (particularly, but not solely, in the poorest rural municipalities).

Note: The results indicated here are after controlling for the size of the conurbation and municipality.

Sources and series: unehistoireduconflitpolitique.fr, figure I.1.

Let's be clear about this: Leaning more heavily on the wealthiest classes than on the others does not mean that the project being proposed to the country is not the most relevant, and conversely, relying more on the working classes is certainly no guarantee that the policy being pursued will be the right one. A voting profile that is too clearly and persistently bourgeois does, however, pose difficulties from a democratic point of view, the main risk being that we become accustomed to the idea that the poorest people are structurally less well placed than the richest to judge the policies that affect them.

The other difficulty with the current situation is that the urban and rural working classes are divided between the left-wing bloc and the national right-wing bloc, which prevents any democratic alternation. To sum up, employees in the service sector (commerce, catering, cleaning, care, etc.) continue to vote left-wing, regardless of their origins, while workers in small towns and villages have swung over to the RN (Rassemblement National, or National Rally).

However, historical experience shows that the tripartition is structurally unstable. It will be difficult for the central liberal bloc to stay in power without broadening its social base in the direction of one or other bloc, probably the right. The most likely and, to a large extent, the most desirable development is the rise to power of a new form of left-right bipolarization, with a left-wing bloc with a broader popular base on the one hand, and a liberal-national bloc on the other, formed by bringing together the most liberal and bourgeois tendencies of the other two blocs. However, this depends on the ability of the left-wing bloc to unite, to deliberate and to democratically internal differences, both on programs and on individuals.

As far as the programmatic basis is concerned, it seems essential to place at the heart of the analysis the very strong feeling of abandonment that has developed since the 1980s and 1990s within the small towns and villages, both in terms of access to public services and to transport, hospital, and educational infrastructures and in terms of the perception of harmful international and European commercial competition orchestrated above all for the benefit of the urbanites. The central point is that the issues at stake are above all socio-economic and require an ambitious and appropriate response in this area. If an appropriate response is not forthcoming, then a continuation of a more or less chaotic tripartition is not out of the question, nor is the perilous transition to a Polish-style bipolarization pitting a social-national bloc against a liberal-progressive bloc, with the attendant risk of tensions escalating and future social and climate challenges going unresolved.

Israel-Palestine: Breaking the Deadlock
October 17, 2023

The atrocities committed during the Hamas terrorist operation, and the ongoing Israeli response in the Gaza Strip, raise once again the question of political solutions to the Israeli-Palestinian conflict and the role that other countries can play in trying to promote constructive developments. Can we still believe in the two-state solution, rendered obsolete in the view of many by the extent of the settlements on the one hand, but also, on the other, by a desire to deny the very existence of Israel and to eliminate its citizens, which has just taken on its most barbaric form with the killings and hostage-takings of the last few days?

Can we still dream of a bi-national state, or is it not time to imagine an original form of confederal structure enabling two sovereign states to one day live in harmony? Such a solution is being mooted more and more often by citizens' movements bringing together Israelis and Palestinians, such as the "A Land for All: Two States, One Homeland" coalition, which has developed innovative and detailed proposals. Too often ignored abroad, these debates deserve to be followed closely.

The Palestinian territories currently have a population of around 5.5 million, including 3.3 million in the West Bank and 2.2 million in Gaza. Israel has a population of just over 9 million, including around 7 million Jewish citizens and 2 million Israeli Arabs. In total, Israel-Palestine has a population of over 14 million, about half of whom are Jews and half Muslims (as well as a small minority of Christians, around 200,000). This is the starting point for the "A Land for All" movement. The two communities are roughly the same size, and each has historical, family, and emotional reasons for considering the land of Israel-Palestine as its own, the land of its hopes and dreams, beyond the arbitrary and tangled borders bequeathed by the military scarring of the past.

Ideally, we would like to imagine a truly bi-national and universalist state, one day bringing together these 14 million inhabitants and granting everyone the same political, social, and economic rights, regardless of their origins, beliefs, or religious practices (the reality and sustainability of which should not be overestimated, especially among the youth of both communities). But before we get there, a long road will have to be travelled to re-establish trust, in the hope that the abject strategy of the terrorists has not destroyed this possibility.

The "A Land for All" coalition initially proposes the coexistence of two states: The current Israeli state and a Palestinian state taking over from the Palestinian Authority established in 1994. The latter, already recognized as a state with observer status at the United Nations since 2012, would exercise full sovereignty over the West Bank and Gaza. What is new is that the two states would be linked by a federal structure guaranteeing freedom of movement and settlement between the two states, similar to the rules applied in the European Union. For example, current Israeli settlers would be able to continue to live and settle in the West Bank, on condition that they respect Palestinian laws, which means that the summary expropriations of Palestinian land carried out over the last few decades would have to stop. Similarly, Palestinians would be free to come to work and settle in Israel, provided they respected the rules in force. In both cases, those choosing to reside in the other state would have the right to vote in local elections.

The authors of the proposal make no secret of the difficulties involved, while showing how they can be overcome. In particular, they explicitly draw inspiration from the European Union, which since 1945 has enabled a century of war and bloodshed between France and Germany to be brought to an end through law and democracy. They also mention the complex case of the Bosnian federation established in 1995. The "A Land for All" coalition also stresses the key role of socio-economic development and the reduction of territorial inequalities. The average salary in Gaza is less than 500 euros, compared with more than 3,000 euros in Israel. The federal entity uniting the two states will have to lay down common rules on labor law, water sharing, and the funding of public, educational, and health infrastructure.

Is there any chance of this happening? That will depend above all on the Israelis and the Palestinians. Having often relied on Hamas in the past to divide and discredit the Palestinians, the Israeli right now seems determined to destroy the terrorist organization. But after that it will not be content to close the lid on the Palestinian territories. It will have to find interlocutors and relaunch a political process.

This is where the rest of the world has a role to play, particularly Europe, which absorbs almost 35% of Israeli exports (compared with 30% for the United States). It is time for the EU to use its trade weapon and make it clear that it will offer rules that are more favorable to a government moving toward a political solution than to a regime playing the decay card. By guaranteeing the Israeli right the same trade rules whatever it does, turning a blind eye to violations of international law and favoring short-term financial interests, the EU has helped to weaken the Israeli left. Yet there is a lively and innovative left in Israel, as in Palestine, particularly among young people. These young people have often found themselves alone in the face of the indifference of governments in both the North and the South, who have made pacts with an increasingly nationalistic and cynical Israeli right. It is now high time to support the side of peace and penalize the side of war.

Taking the BRICS Seriously
November 14, 2023

The war in Gaza threatens to widen the gap between North and South. For many countries in the South, and not only in the Muslim world, the thousands of civilian deaths caused by Israeli bombardments in the Palestinian enclave, 20 years after the tens of thousands of deaths caused by the United States in Iraq, will doubtless embody the West's double standards for a long time to come.

All this is taking place against a backdrop in which the main alliance of so-called emerging countries, the BRICS, has just been strengthened at its Johannesburg summit a few months ago. Initially created in 2009, the BRICS have comprised five countries since 2011: Brazil, Russia, India, China, and South Africa.

Expressed in terms of purchasing power parity, the combined GDP of these five countries will exceed €40,000 billion by 2022, compared with just €30,000 billion for the G7 countries (United States, Canada, Japan, Germany, France, United Kingdom, and Italy), and €120,000 billion on a global scale (an average of just over €1,000 per month for the world's 8 billion people). Differences in average national income per capita remain considerable, of course: almost €3,000 per month in the G7, less than €1,000 per month in the BRICS, and less than

€200 per month in sub-Saharan Africa, according to the latest data from the World Inequality Lab.

In a few words, the BRICS present themselves to the world as the planet's middle class—those who have succeeded, through hard work, in improving their condition, and who have no intention of stopping there.

In 2014, the BRICS created its own development bank. Based in Shanghai, it remains modest in size but could compete with the Bretton Woods institutions (International Monetary Fund and World Bank) in the future if they do not radically reform their voting rights systems to give greater prominence to the countries of the South.

At the Johannesburg summit in August, the BRICS decided to welcome six new members (Saudi Arabia, Argentina, Egypt, United Arab Emirates, Ethiopia, and Iran) from January 1, 2024, reportedly chosen from among some 40 candidate countries.

Let's face it: It's time for Western countries to overcome their arrogance and take the BRICS seriously. It's easy to point out the many inconsistencies and contradictions within what remains a loose, largely informal club. China's political model increasingly resembles a perfect digital dictatorship, and no one wants it any more than Russia's military kleptocracy. At least that guarantees the other leaders that the club won't stick its nose in their business.

The BRICS also include some long-standing electoral democracies, which are certainly experiencing difficulties, but not necessarily more serious than those observed in the West. India has more voters than all the Western countries put together. Turnout was 67% at the last general election in 2019, compared with just 48% in France in 2022, where there has been a sharp drop (unprecedented for two centuries) in the turnout of the poorest communes relative to the richest.

US democracy has also shown all its fragilities in recent decades, from Guantanamo to the assault on Capitol Hill, and has even tended to set a bad example for Brazil's Trumpists.

What can Western countries do to restore their credibility in the South and reduce global fractures? First of all, they must stop giving lessons in justice and democracy to the whole world, even though they are often ready to make pacts with the worst despots and the most dubious fortunes as long as they can make enough money out of it. More generally, Western countries need to formulate concrete proposals to show that they are finally determined to share power and wealth. This requires far-reaching changes to the global political and economic system, whether in terms of the governance of international organizations, the financial system, or the tax system.

In concrete terms, we need to make it clear that the target is a minimum tax on the planet's most prosperous players (multinationals, multimillionaires), with a redistribution of revenues between all countries, according to their population and exposure to climate change.

This is not at all what has been done to date: Minimum taxation concerns only a small number of multinationals; its rate is too low and easily circumvented; and above all, the profits benefit almost exclusively the major countries of the North. The key focus must be redistributing revenues according to the needs of each country, and not according to existing tax bases. Many countries in the South are extremely poor, particularly in Africa, and face such severe difficulties in running their schools, free clinics, and hospitals that such a system would make an enormous difference, even if it were applied to only a small fraction of the revenues collected from the world's multinationals and multi-millionaires.

In *The Ministry for the Future*, US author Kim Stanley Robinson imagines a world in which the transformation of the economic system only occurs after major climatic catastrophes: A heat wave causing millions of deaths in India, and vengeful eco-terrorism from the South shooting down private jets and sinking container ships, all with the covert support of a UN agency despairing of the North's inaction.

Let's hope that competition from the BRICS will encourage the rich countries to grasp the scale of the challenges and share the wealth before it comes to that.

Escaping Anti-Poor Ideology, Protecting Public Service
December 12, 2023

Let's be clear from the outset: The edifying investigation published by *Le Monde* into the intrusive and ubiquitous procedures undergone by thousands of beneficiaries of the Caisse d'Allocations Familiales (CAF), France's welfare agency, poses fundamental issues for the future of social security and public services, in France, Europe, and the rest of the world. By examining thousands of lines of unduly concealed code, meeting vulnerable people and single parents unjustly hounded for imaginary overpayments, the journalists have shown the dramatic consequences of these blind algorithmic practices on everyday lives.

It should also be pointed out that CAF employees are often the first to denounce these practices imposed by their management as well as political leaders. With limited resources, CAF manages not only family allowances but also the active solidarity income (basic income and income supplement for low wages), housing benefits, benefits for single parents or disabled people, childcare benefits, and so on, for a total of nearly 14 million recipients (around half of all French households). The operating costs of the CAF, like those of the health insurance funds and all the social security funds, have always been extremely modest: between 2% and 3% of benefits paid out, depending on the case, compared with 15% to 20% for private insurance companies. This efficiency is a good thing in itself for a public service, provided we don't push too far in this direction.

The problem is that political powers have constantly put pressure on the funds to further reduce these costs. The situation worsened when Nicolas Sarkozy came to power in 2007, emphasizing the need to mercilessly hunt down social security fraud and benefit recipients suspected of ruining the system. Who cares if all the studies show that tax fraud and white-collar tax evasion involve much larger sums? Since it's hard to take it out on the richest, let's take it out on the poorest! This glorification of the "first of the line" and stigmatization of the poorest (deemed incapable of "crossing the street" to find a job and regularly accused of costing "crazy money" to the state) has become even more pronounced with Emmanuel Macron since 2017. Summoned to flush out fraudsters and crunch numbers with reduced human resources, CAF then embarked on the algorithmic drift uncovered by journalists.

The worst thing about this trend is that an anti-poor ideology ends up leading to a general deterioration in the quality of public service. If you haven't experienced this yourself, ask around. For several years now, if you send a message to the CAF on the interface provided for this purpose, the machine tells you that the messages currently being processed are those received three months ago, and that yours will have to wait (six months later, it's still waiting). On the other hand, if you're accused of overpayment, which is sometimes whimsical, you have to pay up straight away, with no possibility of appeal. For those who can afford it, these ubiquitous situations are painful but manageable. For all those whose finances are strained, it's unbearable. Clearly, the CAF does not have the human resources to provide a quality service and treat users correctly, which is extremely painful for everyone involved.

This deterioration in public service can be seen in several areas. For example, with delays of over six months in obtaining identity papers, reimbursement procedures that are still too cumbersome for health insurance and complementary insurance companies, or the extreme opacity of allocation algorithms in higher education, against a backdrop of a shortage of spots and resources in the most sought-after courses.

The right-wing strategy of stigmatizing the poor and the "assisted" as responsible for the country's ills is doubly losing: It weakens the most modest and leads to the degradation of public services for all and the reign of the every-man-for-himself, at the very moment when we need to allocate more resources to provide for the crying needs in health, education, and the environment. The truth is that waste and undue remuneration are to be found in the private sector, not in social funds and public services.

This new anti-poor ideology is all the more worrying as it lies at the heart of current political recompositions. The anti-squatter law adopted at the end of 2022 by a Rassemblement National (RN, far right)–Les Républicains (LR, right-wing)–Renaissance (Macron's party) coalition is the epitome of this. It also shows the dead ends of this approach: We won't solve the housing problems of tens of millions of poorly housed and poorly insulated households by lashing out at the most precarious and weakening all tenants with shortened leases and accelerated evictions.

This question is also an opportunity to fight the RN on the only ground that counts: that of the weaknesses and inconsistencies of its program. The RN's social conversion is an illusion. The party remains deeply imbued with economic liberalism, as demonstrated by its desire to abolish the real estate wealth tax, in the same way that former RN president Jean-Marie Le Pen wanted to abolish the income tax in the 1980s. It's high time to move away from the current obsession with identity and put socio-economic issues back at the heart of public debate.

Rethinking Europe after Delors
January 15, 2024

With the death of Jacques Delors, president of the European Commission from 1985 to 1995, a chapter in European history has ended. The time has come to take critical stock of this decisive period and to draw lessons for the future, a few months ahead of the European elections of June 2024.

To say that the Europe we know today was shaped during this period would be an understatement, with the 1986 Single European Act (allowing for the free movement of goods and services), the 1988 European Directive on the liberalization of capital flows, and the 1992 Maastricht Treaty. In particular, it was the Maastricht Treaty, narrowly adopted by French voters in September 1992 (51% yes), that transformed the former European Economic Community (EEC, established in 1957 by the Treaty of Rome) into the European Union (EU) and endowed it with a single currency. As planned in 1992, the euro came into effect in 1999 for companies and in 2002 for individuals. The 2005 Treaty establishing a Constitution for Europe—rejected in France in a referendum (55% voted no) and then adopted by Parliament after a few minor changes in the form of the Lisbon Treaty in 2007—basically confined itself to consolidating the crucial decisions made between 1986 and 1992 and constitutionalizing the principles of free competition and free movement, without any major innovations. The 2012 European Fiscal Compact tightened up the Maastricht criteria on debt and deficits laid down in 1992, again without any significant innovation.

To understand what was at stake in the decisive European negotiations between 1985 and 1995, the primary reference work remains *Capital Rules*, by Rawi Abdelal.[1] Based on dozens of in-depth interviews with the main political players and senior European officials at the time, in particular Jacques Delors and Pascal Lamy, Abdelal skillfully analyzes the visions of the future and the negotiating leeway of both sides. In a nutshell, the French Socialists were gambling that the creation of the euro and the European Central Bank (ECB), a powerful federal institution that makes decisions by majority vote, would eventually enable the establishment of a European public power capable of regulating economic forces more effectively than the French left-wing unity government that emerged from the 1981 elections. To achieve this result, they agreed to the central demand of the German Christian Democrats, who advocated absolute liberalization of capital flows, without any public regulation, and in particular without any common taxation. This was a crucial issue largely neglected by French President François Mitterrand and Delors during the negotiations. The foundations of the compromise were laid.

Thirty years on, the outcome of these radical innovations is inevitably mixed. On the one hand, the ECB played a central role in preventing a widespread collapse in the wake of the 2008 financial crisis and the Covid-19 pandemic. After some

---

[1] Rawi Abdelal, *Capital Rules: The Construction of Global Finance* (Harvard University Press, 2007).

initial blunders during the Greek crisis and the unnecessary austerity relapse of 2012–2013, majority decision-making enabled the ECB to override national (notably German) vetoes and quickly and efficiently mobilize considerable sums to stabilize the European economy and reduce interest rate differentials within the eurozone. No one knows what would have happened without the single currency, and it has to be said that the Nordic countries that remained outside the euro did not fare so badly. The truth remains that no credible political player is proposing a return to the franc.

Conversely, it is well understood that money creation alone cannot solve all problems. Moreover, central bankers have been far more willing to save banks and bankers than to invest in education, health, and the climate. In this way, they have contributed to the increasing concentration of wealth, with the richest benefiting from the swelling of the stock market and real estate assets made possible by share buybacks and public money, while the savings of the poorest are being wiped out by ongoing inflation. The European rules on the free movement of capital laid down in 1992 have proved so extreme and destabilizing that even the IMF decided after the Asian crisis of 1997 and then 2008 to reintroduce some form of capital controls for short-term flows.

The new European rules have also played a major role in exacerbating tax dumping: endless corporate tax cuts, the unprecedented development of tax havens, and the structural under-taxation of billionaires and multi-millionaires. Politically, the 1992 and 2005 referendums contributed significantly to alienating a proportion of the working class from the ballot box and from the left. The "no" vote in 2005 in France thus was the best predictor of the vote for the far-right Rassemblement National party in 2022, particularly in medium-sized towns affected by deindustrialization.

What can we do about this complex legacy? First, we need to propose to our partners that they set up a strong core within the EU capable of making majority decisions on budgetary, fiscal, and environmental matters. Even if this European Parliamentary Union does not see the light of day in the immediate future, it remains the central objective. Second, until a compromise is reached, it will undoubtedly be essential to adopt substantial unilateral measures in the face of intra-European and extra-European fiscal, social, and environmental dumping. This will give rise to complex crises that can be overcome if we maintain a consistent internationalist course, and are probably inevitable if we want to break out of the current deadlock.

Peasants, the Most Unequal of Professions
February 13, 2024

The French and European agricultural crisis has demonstrated that no sustainable development trajectory is possible without a drastic reduction in the social inequalities and glaring injustices of our economic system. Instead, the public authorities in Paris and Brussels are embarking on an old-fashioned headlong rush to relaunch pesticides and pollution, without giving themselves the means to tackle injustices and liberal dogmas. This is all the more ill-adapted given that the farming world today is the most unequal of all professional universes. No viable solution can be found without starting from this basic material reality.

Let's take a step back. In recent weeks, French public opinion has been struck by a widely shared statistic: The average annual income of farmers reached €56,014 in 2022, a much higher level than sometimes imagined. The data, compiled by the Ministry of Agriculture's statistical services for European comparison purposes, is also available at the most detailed level, by farm type and by income decile. In order to interpret it correctly, however, several points must be taken into account. First of all, the study excluded some of the smallest farms. The

Ministry specified that 95% of surfaces and 99% of production were covered. However, between 10% and 20% of farmers were excluded, depending on the sector.

Secondly, and most importantly, consider the concept of income used in the study. It is the average annual income per full-time farmer, after deduction of all operating expenses, including financial charges (loan interest) and equipment amortization charges, but before deduction of income tax and all social security contributions. This largely explains why the average income of €56,014 is so high.

If the average remuneration per employee (full-time equivalent) in France in 2022 is calculated, including all employee social contributions (deducted from gross salary) and employer social contributions (paid by employers in addition to gross salary), then we also arrive at an average of around €60,000 per year, close to that of farmers, or a little higher. With an equivalent concept, the average income of doctors reaches €120,000 per year (90,000 for general practitioners, 150,000 for specialists).

It's true that farmers, like all self-employed people, have much lower social security contributions than salaried employees, so their average disposable income after deducting contributions is significantly higher. But these lower contributions also translate into lower pensions and other social entitlements, forcing farmers to save more to compensate.

Even more so than doctors and other self-employed people, farmers are also forced to tie up extremely large amounts of capital, which they can in principle sell when they retire, although this operation is not without risk. In the end, the farmers' average income of €56,014 is in no way exorbitant when compared with the rest of the country's working population.

What really sets farmers apart, however, is the extreme inequality of pay distribution around this average. According to the data available, farmers even appear to be the most unequal of all professions in France today.

Generally speaking, pay inequalities within self-employed occupations are significantly higher than within salaried occupations, owing in particular to difficulties in accessing capital and equipment. And among the self-employed, income inequalities are significantly higher among farmers than in other professions, such as shopkeepers, restaurateurs, bakers, transport, and construction. In concrete terms, for an average income of €56,014, according to the statistical services of the Ministry of Agriculture, 25% of farmers exceed €90,000 and 10% exceed €150,000. Incomes of several hundred thousand euros a year are not uncommon, particularly among the current leaders of the FNSEA farmers' union, who often combine their activity as a farmer-manager with that of a shareholder in the agro-industry.

At the other end of the scale, the lowest-paid 10% of farmers earn less than €15,000, in many cases well below the minimum hourly wage, given the long working days. There are also considerable differences between farm categories, with average incomes ranging from €19,819 for cattle and goat farmers to €124,409 for pig farmers, whose incomes have varied widely but have risen sharply over the last 30 years.

What can we conclude from all this? Firstly, that global solutions make no sense. Abolishing the tax on agricultural diesel or reintroducing pesticides will obviously bring in much more money for those already earning €150,000 than for those on €15,000. Secondly, that it makes no sense to respond to competition from foreign pesticides by reducing standards on French production.

A much better solution would be to immediately introduce safeguard measures aimed at making the imports concerned pay for the undue benefit they derive from non-compliance with French standards. It is by tackling the inequalities of the farming world and the challenges of organic farming head-on that we will emerge from the current crisis.

When the German Left Was Expropriating Princes
March 19, 2024

Just over a century ago, in the spring of 1924, the German left launched an uphill battle to redistribute the wealth of the Hohenzollerns, the ruling family who had lost power with the abdication of Wilhelm II and the creation of the Weimar Republic in 1919. Rich in lessons for today, this little-known episode deserves to be remembered. It illustrates the ability of elites to use the language of the law to perpetuate their privileges, regardless of the scale of their wealth or the importance of collective needs. Yesterday, it was the reconstruction of European societies ravaged by war; today, it is the new social and climatic challenges.

The episode is all the more interesting given that the Weimar Constitution is considered one of the most advanced in social and democratic terms. In particular, both the Constitution of 1919 and the Basic Law of 1949 adopted an innovative definition of property as a social right, rather than a strictly individual and unlimited right, irrespective of material needs and the social groups concerned. The 1919 text stipulated that the law should determine the system of real estate ownership and the distribution of land on the basis of social objectives, such as ensuring "a healthful habitation and to all German families" and "homesteads for living and working that are suitable to their needs" (article 155). Adopted against a backdrop of near insurrection, the text led to major land redistribution and new social and trade union rights.

The 1949 text affirms that property rights are legitimate only insofar as they "serve the general welfare" (article 14). It explicitly mentions that the socialization of the means of production and the redefinition of the property regime fall within the domain of the law (article 15). The terms used open up the possibility of structural reforms such as co-determination. The 1951 law decided that employee representatives should have 50% of the seats on the governing bodies (boards of directors or supervisory boards) of the major steel and coal companies, irrespective of any shareholding. The 1952 law extended the system to all sectors of activity. The 1976 law established the current system, with one-third of seats for employees in companies with between 500 and 2,000 employees, and one-half of seats for those with more than 2,000 employees.

It was also against this backdrop that, in 1952, the German Parliament adopted an ambitious *Lastenausgleich* ("burden equalization") scheme, consisting of a levy of up to 50% on the highest financial, professional, and real estate assets (whatever their nature). The system raised considerable sums (around 60% of German national income in 1952, with payments spread over 30 years). This made it possible to finance substantial compensation for the small and medium-sized estates slashed by the destruction and monetary reform of 1948 (1 new mark replaced 100 old marks, making it possible to get rid of the immense public debt without inflation), and to render politically acceptable this essential measure for injecting new life into public finances. By all accounts, this revolutionary system played a central role in rebuilding the country on the basis of a new social and democratic contract.[1]

However, in the context of the political struggles of 1924–1926, this constitutional modernity was not enough. In Austria, the Habsburgs' imperial estates had become community property without compensation. In Germany, however, the Hohenzollerns managed to retain their properties (over 100,000 hectares of land, a dozen castles, works of art galore, etc.). No federal redistribution measures were adopted. Several rulings in 1924–1925 also invalidated decisions by regional governments to limit private use and open up public access to castles and works of art. Following the hyperinflation of 1923, the Hohenzollern princes went so far as to demand an increase in their pensions, while the country was on its knees.

The Communists of the KPD, eventually followed by the Social Democrats (SPD), introduced a bill to expropriate the princes for the benefit of the poorest. They gathered over 12 million signatures in 1925, in what remains to this day the largest petition in German history. The law was about to be passed, but the vagueness of the constitutional wording on compensation allowed President Hindenburg to demand a constitutional revision beforehand. The June 1926 referendum attracted 16 million voters (90% in favor of expropriation). However, the turnout was slightly below the 50% threshold required to amend the constitution.

By calling for abstention and denouncing the risks that a Communist victory would eventually pose for small and

---

[1] Michael L. Hughes, *Shouldering the Burdens of Defeat: West Germany and the Reconstruction of Social Justice* (University of North Carolina Press, 1999).

medium-sized property owners, the German right and large landowners (very influential in the east of the country), allied with the center and the Nazis (who opposed class struggle and advocated expropriating Jews who had arrived in the country since 1914), succeeded in blocking the process and preventing the left-wing union that could then have been put in place.

The episode is fundamental, as it illustrates the importance of constitutional battles in the historic march toward equality. It's a process that is still ongoing and one that will undoubtedly see new developments in the decades to come.

Should Ukraine Join the EU?
April 16, 2024

Is Ukraine's possible entry into the European Union a good idea? Yes, but on the condition that the European project is also redefined at the same time. In short, it should be an opportunity to redefine the EU as a political community serving the rule of law and democratic pluralism; and to break away from the economic religion of free trade and competition as the solution to all problems, which has dominated the construction of Europe for several decades.

The defense of Ukraine from Russia is vitally important, and this is first and foremost for political and democratic reasons. Unlike its Russian neighbor, Ukraine respects the principles of electoral democracy, democratic alternation, separation of powers, and peaceful conflict resolution.

Ukraine's entry into the EU must be an opportunity to define strict standards guaranteeing pluralism in all its forms, both in terms of organizing electoral life (with legislation that is, finally, ambitious in terms of campaign and party financing) and regulating the media (with solid guarantees of independence for editorial teams, and real power-sharing between journalists, citizens, and shareholders, both public and private).

Europe likes to present itself to the world as a near-perfect democratic club, a guiding beacon for the world. However, while the practice of electoral democracy is, in some respects, more advanced here than in other parts of the world, its institutional foundations remain no less fragile and incomplete. The challenge is not only to defend transparency in Kyiv and challenge Ukrainian oligarchs' political stranglehold on elections and the media; but also to reduce the power of French, German, Italian, Polish, or Maltese oligarchs, and to promote new, more democratic, pluralist, and egalitarian forms of political participation throughout the EU—protected from the powers of money and private interests.

The adoption of more ambitious European democratic standards must also be an opportunity to break away from the free-trade and competition-oriented economic religion that has stood in for a European philosophy since the Single European Act of 1986 and the Maastricht Treaty of 1992.

In other words, to prevent Ukraine's entry into the EU from causing further social and environmental damage, notably as a result of exacerbated competition in the agricultural and industrial sectors, it is crucial to act simultaneously on two fronts. Firstly, it is imperative to do everything we can to establish, as quickly as possible, a core group of countries within the enlarged EU that would be prepared to adopt stricter social, fiscal, and environmental standards by a majority vote. This could be achieved, for example, through the draft treaty on the democratization of Europe, with the creation of a European Assembly drawn from the national parliaments of those countries that are ready for closer integration. Other solutions are conceivable, on the condition that they can be implemented by a small number of willing countries, without any possibility of being blocked by the other countries.

Then, while awaiting the establishment of such a core group—and also in order to sustainably complement its actions—it is vital that each country provides itself with the means to set conditions on continuing trade with other countries, including its European partners.

A particularly clear example concerns tax dumping. The problem with the minimum corporate tax rate of 15% negotiated at the OECD and EU level, apart from the fact that it contains numerous loopholes, is that it is ridiculously low. Given the requirement for unanimity in the current rules, this is unlikely to change in the foreseeable future.

The simplest way to break the deadlock on the situation is to take unilateral action. For example, if a country like France considered that the appropriate rate of tax on profits was, let's say, 30%, then it could very well decide that imports from countries where the rate was lower would have to pay the difference when the goods and services are sold on its territory. As the EU Tax Observatory has shown, such a measure would bring in €39 billion in revenue for France—in other words, considerable resources for investment in healthcare, education, and transport.

Advocates of generalized dumping will cry "protectionism," but the truth is that this is totally different: It's simply a matter of making companies that export goods and services to France pay the same rate as that paid by producers based in this territory, which should have long been considered a minimum condition for fair trade.

The same logic could be applied to health standards or carbon emissions. In this respect, it should be remembered that the Carbon Border Adjustment Mechanism, adopted by the EU in 2022, has been calibrated to bring in barely €14 billion a year from now until 2027, or the equivalent of customs duties that are lower than 0.5% of the total amount of non-European imports entering the EU (and just over 2% of the total amount of Chinese imports). To be clear: To have a significant impact on trade flows between Europe and the rest of the world, the sums involved would have to be multiplied by 10 or 20. Here again, unilateral action is the only way out of the current deadlock.

It is by re-establishing our social and economic room for maneuver that we will be able to convince public opinion of the advisability of a new European enlargement, based on shared democratic values—and not on a liberal economic religion that benefits the richest and drives the middle and working class ever further away from the European ideal.

For a Binational Israeli-Palestinian State
May 14, 2024

Can a two-state solution in Israel-Palestine still be reached, and under what conditions would it be viable? A word of optimism first: There are many citizen peace movements in both Israel and Palestine who tenaciously and imaginatively advocate peaceful, democratic solutions. Unfortunately, these groups are in a minority, and without powerful external support they have little chances of prevailing.

To break the stalemate, it's time for the European Union and the United States, which between them absorb almost 70% of Israeli exports, to match rhetoric with action. If Western governments truly support the two-state solution, then sanctions must be imposed on the Israeli government, which is openly trampling all peaceful prospects by pursuing settlement and repression and opposing recognition of the Palestinian state.

In other words, military aid must stop, and above all the United States and Europe must hit Netanyahu and his allies in the wallet. This means introducing trade and financial sanctions, gradually increasing to dissuasive levels. The academic boycott of universities that has been started will not be enough, and may even prove counter-productive. It is often on campuses that the main opponents of the Israeli right can be found, and in many cases the right will be delighted to weaken them and cut them off from the outside world. At the same time as imposing sanctions on Israel, Europe, and the United States must put in place implacable and dissuasive sanctions against Hamas and its external supporters, and decisively strengthen representative and democratic Palestinian organizations.

This major external involvement, which ideally should bring together Western countries and a coalition of countries from the South, is all the more essential given that no two-state solution will be possible without a strong confederate structure—a form of Israeli-Palestinian Union, similar to the European Union—covering both states and guaranteeing a certain number of fundamental rights. The two territories and populations are deeply intertwined, due to the scale of Jewish settlement in the West Bank, the large number of Palestinian workers employed in Israel with family ties to Israeli Arabs, and the non-contiguity of the Palestinian territories. To begin with, the Israeli-Palestinian Union would have to guarantee freedom of movement and establish a minimum basis of social and political rights for Israelis living or working in Palestine, as well as for Palestinians living or working in Israel. One of the most successful projects along these lines is that developed by the remarkable Israeli-Palestinian citizen movement "A Land for All," too often ignored abroad.

Eventually, this confederate structure could become a genuine bi-national Israeli-Palestinian state treating all its citizens equally, regardless of their origins, beliefs, or religions. But for the process to get underway, extremely strong external pressure will be essential, backed up by substantial financial resources (but well within the reach of Europe or the United States) and a multinational force to enforce the agreement and disarm Hamas and extremist groups on both sides.

The challenges may seem immense, but what's the alternative? To wait peacefully for the open slaughter of Palestinian civilians to reach 40,000 dead, then 50,000 dead, then 100,000 dead? The moral and political cost of Western inaction is exorbitant. It can be explained first and foremost by the navel-gazing of European and American societies, too preoccupied with their own divisions to take any real interest in constructive solutions in Israel-Palestine. There is, of course, the old anti-Semitism, never extinguished and always ready to be rekindled, based on ignorance and misunderstanding of the other. Every Jew is accused of complicity with Israeli generals, just as stupidly as every Muslim is suspected of complicity with jihadists.

What's also new is the shameful instrumentalization of the fight against anti-Semitism. On the right and now also in the center, pro-Palestinian mobilizations are immediately branded as anti-Semitic—even by notorious anti-Semites—and associated with an imaginary Islamo-Leftism, without any concern for the reality of the speeches and proposals. The fact that there are provocateurs ready to play with fire in all camps is obvious, but it's always possible to distance oneself clearly and focus on what counts. Unfortunately, fear (not to say hatred) of Islam and European Muslims sometimes seems to block any calm reflection. Accusations of anti-Semitism allow us to clear our consciences while turning a blind eye to the ongoing massacres.

In the United States, the Muslim minority is smaller than in Europe and causes less tension, but the political reflexes are the same, with the added bonus of a Messianic, semi-delusional mobilization of evangelical Christians in favor of Israel. Conversely, a large proportion of Jewish students and secular Jews of all ages are now mobilizing across the Atlantic for Palestinian rights. This is the main reason for hope. On both sides of the Atlantic, young people are rejecting old divisions as well as new hatreds. They see clearly that what is at stake in Israel-Palestine is the possibility of living together beyond our origins. It is on this hope that we must build the future.

For a Geopolitical Europe, neither Naive nor Militaristic
June 8, 2024

Unsurprisingly, the debates leading up to the 2024 European elections were marked by geopolitical issues: the wars in Ukraine and Gaza and growing tensions between the West and the China-Russia bloc, which intends to increase its influence in the South and increase members of the BRICS+ group (Brazil, Russia, India, China, South Africa, Egypt, United Arab Emirates, Ethiopia, and Iran). Some say the cause is clear: Europe's future will be more kaki-oriented. Faced with the Russian threat, the European Union (EU) has no choice but to flex its military muscles and massively increase the budget for its militaries, which some would like to see rise from the current 1.5–2% of national income to 3%, or even 4%.

However, there is nothing to suggest that such a prospect is realistic or even desirable. Firstly, because Western military budgets are already considerable and would benefit from being better mobilized. Secondly, because Europe would be better advised to put its wealth and power at the service of social, educational, scientific, and climatic objectives. Finally, and above all, because Europe must try to influence other countries through economic and financial sanctions, the law and social justice, rather than military, means. Instead of falling into the temptations of defense-focused geopolitics, Europe must invent a social, economic, and climate geopolitics.

Let's start by recalling that the North Atlantic Treaty Organization (NATO) countries are collectively far more powerful economically and militarily than Russia. Their combined gross domestic product is ten times higher, and their air capabilities five times greater. The problem is that NATO has decided to let Russia bomb Ukrainian territory as much as it likes, including massacring civilians and destroying homes and energy infrastructure.

With the air capabilities at its disposal, NATO could decide to impose a no-fly zone over Ukraine. As long as the aim is to defend Ukrainian territory, and in no way to attack Russian territory, such a mobilization of both NATO's human and material resources would be legitimate. Lending a few aircraft or anti-aircraft batteries to Ukraine won't be enough, as it takes years to train qualified pilots and personnel. In any case, Ukraine will remain at a massive demographic disadvantage to Russia.

The strategic decision to intervene directly is admittedly a difficult one: It would mean nothing less than defending Ukraine, as NATO would have to do in the event of aggression by one of its members. But the fact is, it would be just as difficult if NATO had ten times as many aircraft as Russia. Once this red line has been drawn, Western countries could also open the door to legitimate, democratic political processes in the disputed territories of Crimea and Donbas.

To sum up, the challenge is not to massively increase Western military budgets—they are already considerable—but to know how to mobilize them. The issue is not financial, but decision-making. As far as financial resources are concerned, it would be in the EU's interest to invest additional resources in health, training at all levels, scientific research, transport and energy infrastructures, housing, energy-efficient renovation of buildings, sustainable agriculture, and decarbonization, with social justice and for the benefit of the middle and working classes.

Due to the social struggles of the twentieth century, Europe already has the best primary and secondary healthcare and education system in the world, far ahead of the United States. In the twenty-first century, Europe must also have the best universities on the planet. France, Germany, and their European allies have the financial resources needed to finally make such a choice, and yet are doing nothing, out of ideology and ignorance. The situation is particularly absurd in France, where spending per student has fallen by 15% over the last ten years, contrary to all historical trends.

To increase its influence in the world, Europe must first and foremost promote and perfect its social, economic, and democratic model. To influence other countries, it must not rely on gunboat diplomacy (which was amply used and abused from 1492 to 1962) and the superiority of its military means (except for strictly defensive purposes), but on tools consistent with its social model. There are, of course, the classic trade sanctions, which need to be reintroduced into the political repertoire as a matter of urgency.

It makes little sense to insist on absolute free trade with China, while massively importing its carbon emissions (and the EU's tiny border carbon tax won't change a thing) or watching the Chinese regime destroy before our very eyes electoral democracy in Hong Kong. We also need to develop new types of financial sanctions targeted at the elites of certain countries, through the implementation of a veritable global financial registry and payment system exclusion measures, as already practiced in the United States to enforce its anti-money laundering measures.

Europe urgently needs to seize the €200 billion of Russian public assets located on its territory and do the same with Russian private assets (between €500 billion and €1 trillion, depending on estimates). By stubbornly insisting on the sanctity of financial orthodoxy and oligarchic ownership (however ill-gotten it may be, as long as it's profitable), an attitude that did not save the Belle Epoque (1870–1914) from the disaster that followed, Europe is turning its back on its own history, undermining its international moral credibility and condemning itself to being a geopolitical miniature.

Rebuilding the Left
July 13, 2024

Despite the relative majority obtained by the Nouveau Front Populaire (NFP, left-wing alliance) in legislative elections, the French political landscape remains marked by divisions and uncertainty. Let's be clear: The left's gains in votes and seats are actually very limited, and reflect insufficient work on both policy and structure. It is only by resolutely tackling these shortcomings that the left-wing parties will be able to get through the period of turbulence and minority governments that lies ahead, and one day obtain the absolute majority that will enable them to govern the country on a long-term basis.

The policy platform adopted by the NFP a few days after the dissolution of the Assemblée Nationale did have the immense merit, compared to the others, of identifying where to find the resources to invest in the future: health, training, research, transport, and energy infrastructures. These essential investments are going to increase sharply, and there are two ways of funding them. Either we proclaim that we are entering a new cycle of increasing socialization of wealth, driven by tax increases on the wealthiest, as proposed by the NFP, or, out of ideology, we refuse any tax increase at all, thereby putting ourselves in the hands of private funding, synonymous with inequality of access and a more than dubious collective efficiency. Boosted by staggering private costs, health spending is approaching 20% of GDP in the United States, even as the indicators are disastrous.

However, the amounts mentioned by the NFP may have frightened some: Around 100 billion euros in new levies and expenditure over the next three years, or 4% of GDP. In the long term, these amounts are not excessive: Tax revenues have risen in Western and Nordic Europe from less than 10% of national income before 1914 to 40–50% since 1980–1990, and it is this rise of the social state (education, health, public services, social protection, to name a few) that has enabled unprecedented growth in productivity and living standards, whatever the conservatives of any era may have said.

The fact remains, however, that there is considerable uncertainty as to the timetable and order of priorities for a left-wing government coming to power. While the demand for social justice is strong in the country, the mobilization of new resources remains a fragile process from which citizens can withdraw their support at any time. In concrete terms, until it has been incontrovertibly demonstrated that billionaires and multinationals are finally being made to contribute, it is unthinkable to ask anyone else to make an additional effort. The NFP policy platform remains too vague on this crucial point.

This is all the more problematic given that the left-wing governments of recent decades, lacking a sufficiently precise program and sufficiently strong collective ownership of it, always found themselves caving in to the lobbies as soon as they came to power, for example by exempting so-called professional assets and virtually all the largest fortunes from the ISF (wealth tax), resulting in revenues ridiculously low compared to what they could and should be. To avoid repeating these mistakes, we need to involve civil society and trade unions in defending these revenues and the social investments that go with them. On these and other issues, slogans are no substitute for hard work and collective mobilization.

Similar difficulties are encountered on the question of pensions. It doesn't make much sense to adopt the slogan of retirement for all at 62, or even 60, when everyone knows that the length of contribution to obtain a full pension is also a requirement in the French system. A slogan like "42 years of contribution for all" would be better understood by the country, and would make it clear that people with higher education will not retire before 65 or 67, while insisting on the unacceptable injustice of Emmanuel Macron's reform raising the retirement age to 64, which forces, for example, those who started working at 20 to contribute 44 years.

There are many examples. It's all very well to announce the abolition of Parcoursup, the portal that has handled university applications since 2018, but it would have been even better to describe precisely the alternative, fairer, and more transparent system that would replace it. It's all very well to denounce Vincent Bolloré's media, but it would be even better to commit to an ambitious law to democratize the media and challenge the all-powerful shareholders.

There's also the proposal to give employee representatives one-third of the seats on company boards. This is the most far-reaching and genuinely social-democratic reform in the NFP program, but it would benefit from an even broader framework. To enable the redistribution of economic power, we would need to go as far as 50% of seats in large companies, while capping the voting rights of the largest shareholders and committing to a genuine redistribution of wealth. Rather than wallowing in rhetorical radicalism, it's time for the left to get back to describing the alternative economic system to which it aspires, while recognizing that things will happen in stages.

On all these questions, only collective work will allow progress, which requires the creation of a genuine democratic federation of the left capable of organizing deliberation and settling disputes. We're a long way from that: In recent years, La France Insoumise has constantly sought to impose its authoritarian hegemony on the left, in the manner of Socialists of yesteryear, only worse, given the refusal of any voting procedure on the part of LFI's leaders. But the left-wing electorate is not fooled: It knows full well that the exercise of power requires above all humility, deliberation, and collective work. It's time to respond to this aspiration.

Europe Must Invest: Draghi Is Right
September 17, 2024

Let's face it: The report on Europe's competitiveness and future submitted by Mario Draghi to the European Commission is heading in the right direction. For the former European Central Bank president, Europe needs to make 800 billion euros of additional investments per year in the future—the equivalent of 5% of the European Union's (EU) GDP—or around three times the Marshall Plan (between 1% and 2% of GDP in annual investments in the post-war period). This would enable the continent to return to the investment levels of the 1960s and 1970s. To achieve this, the report proposes recourse to EU borrowing, as was done with the €750 billion recovery plan adopted in 2020 to cope with Covid-19. Except that the aim now is to raise such sums each year for sustained investment in the future (particularly in research and new technologies), and not to fund an exceptional response to a pandemic. If Europe proves incapable of making these investments, then the continent will enter a "slow agony" in the face of the United States and China, the report warns.

One may disagree with Draghi on several key points, not least of which is the precise composition of the investment in question. Nevertheless, this report has the immense merit of challenging the dogma of fiscal austerity. According to some, in Germany but also in France, European countries should repent for their past deficits and enter a long phase of primary surpluses in their public accounts, in other words, a phase in which taxpayers should pay much more in taxes than they receive in expenditure, to urgently repay the interest on the debt and the principal.

In reality, this austerity dogma is based on economic nonsense. Firstly, because real interest rates (net of inflation) have fallen to historically extremely low levels in Europe and the United States over the last 20 years: Less than 1% or 2%, and sometimes even negative levels. This reflects a situation where there is a huge windfall of little-used or misused savings in Europe and worldwide, ready to pour into Western financial systems with virtually no yield. In such a situation, it is the role of public authorities to mobilize these sums and invest them in training, healthcare, research and new technologies, major energy and transport infrastructures, thermal renovation of buildings, and so on.

As for the level of public debt, it is indeed very high, but not unprecedented. It is close to that observed in France in 1789 (around one year's national income), and significantly lower than that seen in the United Kingdom after the Napoleonic Wars and in the nineteenth century (two years' national income) and in all Western countries after the two World Wars (between two and three years).

Yet history shows that such high levels of debt cannot be dealt with using ordinary methods. Exceptional measures are needed, such as levies on the highest private assets, like those successfully applied in Germany and Japan in the postwar period. When real interest rates rise again, we'll have to do the same, by taxing multi-millionaires and billionaires. Some will argue that this is impossible, but in reality it's a simple book-entry transfer. The same cannot be said of global warming, public health, or training challenges, which cannot be solved with the stroke of a pen.

If we now examine the details of the Draghi report's proposals, there is obviously much to criticize, and that's a good thing. Once the principle that Europe needs to invest massively has been accepted, it's healthy for different visions to be expressed as to the type of development model and welfare indicators we want to promote. In this case, Draghi's approach is technophile, mercantile, and consumerist. He emphasizes large-scale public subsidies for private investment in digital technology, artificial intelligence, and the environment. However, there is every reason to believe that Europe should seize the opportunity to develop other modes of governance and avoid, once again, giving full power to large private capitalist groups to manage our data, our energy sources, or our transport networks.

Draghi also considers public investment, for example in research and higher education, but in an overly elitist and restrictive way. He proposes that the European Research Council should finance universities directly (and not just individual research projects), which would be a very good thing. Unfortunately, the report proposes to focus solely on a few poles of excellence in major metropolises, which would be economically dangerous and politically unacceptable. Public health and hospitals are almost entirely absent from the report.

Generally speaking, for such an investment plan to be adopted, it is essential that disadvantaged territories and the most disadvantaged regions—including, for example, in Germany—benefit from massive and visible resources. If

France, Germany, Italy, and Spain, which together account for three-quarters of the eurozone's population and GDP, can agree on a balanced, socially and territorially inclusive compromise, then it will be possible to move forward without waiting for unanimity, relying on a core group of countries (as envisaged in the Draghi report). This is the debate that Europe must now engage in.

How to Tax Billionaires
October 15, 2024

The tax debates currently underway in France and the discussions which took place at the 2024 G20 summit demonstrate that the issue of tax justice and the taxation of billionaires is not about to disappear from the public debate. There's a simple reason for this: The sums amassed by the world's wealthiest individuals over the last few decades are quite simply gigantic. Those who consider this a secondary or symbolic issue should take a look at the numbers. In France, the combined wealth of the 500 largest fortunes has grown by €1 trillion since 2010, rising from €200 billion to €1.2 trillion. In other words, all it would take is a one-time tax of 10% on this €1 trillion increase to bring in €100 billion, which is equal to all of the budget cuts the government is planning for the next three years. A one-time tax of 20%, which would remain very moderate, would bring in €200 billion and allow as much additional spending.

Yet some people continue to reject this debate, and their arguments need to be carefully examined. The first is that these immense private fortunes are merely theoretical and don't really exist. While it's true that they often appear as numbers on a screen, just like public debt or salaries paid into bank accounts, these figures have a very real impact on people's lives and influence the power dynamics between social classes and public authorities. Concretely, how would billionaires pay this 10% tax on their wealth increase? If they don't make enough profit in a year, they'll have to sell some of their shares—say, 10% of their portfolio. If finding a buyer is challenging, the government could accept these shares as payment for taxes. If necessary, it could then sell these shares through various methods, such as offering employees to purchase them, which would increase their stake in the company. In all cases, net public debt will be reduced accordingly.

The second argument often heard is that modern governments are too weak to impose anything on billionaires. With globalization and the free movement of capital, billionaires can simply relocate to more favorable jurisdictions, making any expected tax revenue vanish. While this argument may seem convincing to some, in reality, it is hypocritical and weak. Firstly, it was governments that set up the free movement of capital, upheld by a sophisticated legal system backed by the public courts, which could potentially be changed. Secondly, this argument reflects an abandonment of sovereignty, particularly from political leaders who frequently discuss the need to restore government authority but often find it easier to exercise their authority over the poor than over the powerful.

Finally, and most importantly, this defeatist rhetoric overlooks the fact that governments still have room to maneuver, including the ability to act independently. For example, when the United States threatened to withdraw Swiss bank licenses, Bern put an end to its banking secrecy. Similarly, across the Atlantic in the United States, taxpayers are taxed according to their nationality, even if they live abroad. If someone wants to give up their US passport, an option not without risk, there's nothing to prevent the government from continuing to tax them, as long as their wealth was accumulated in the United States, or even more simply, if they continue to use the dollar.

France is a smaller country, but it also has considerable leverage. France, for example, could impose an exceptional wealth tax based on the number of years spent in France. A taxpayer who has been a resident in Switzerland for one year after spending 50 years in France would, for example, continue to pay 50/51sts of the tax required by a French resident. Those who refuse to pay would be outlawed and could face legal penalties.

The final argument against taxing billionaires is that it would be illegal or unconstitutional. This is nothing new: Throughout history, the powerful have often invoked legal language to preserve their privileges. However, there is nothing in the Constitution that prevents the implementation of an exceptional tax on the wealth of billionaires, or more generally, wealth taxation, which is a valid indicator of citizens' ability to pay taxes just as much as income. In fact, this is why a comprehensive system of inheritance and property taxes was established in 1789, and why an exceptional wealth tax was introduced in 1945. The fact that some constitutional judges ignore this and sometimes try to use their position to impose their partisan preferences doesn't change a thing: This is fundamentally a political debate, not a legal one.

Other solutions are possible, such as Prime Minister Michel Barnier's tax on incomes over €500,000. However, this tax will bring in €2 billion compared to the €100 billion that could be raised from a 10% tax on the wealth of billionaires. The reason for this disparity is that billionaires' income constitutes only a tiny fraction of their overall wealth, meaning they would effectively pay very little under the Barnier tax. Consequently, it is the most modest who will bear the brunt of the Barnier budget and the cuts to public services. This strategy leads us straight to the wall: We can't effectively address today's social and climate challenges if we don't start by taxing the wealthiest in a clear and significant way.

Unite France and Germany to Save Europe
November 12, 2024

In the wake of Donald Trump's victory, Europe can no longer content itself with declarations of intent. It urgently needs to pull itself together and regain a grip on world affairs, without any illusions about what will come from the United States. The crux of the matter is that it is impossible to face up to the socio-economic, climatic, and geopolitical challenges that are shaking the planet as long as the European Union (EU) makes its decisions by the unanimous vote of all 27 member states. Unfortunately, this is currently the case for all major decisions, particularly those with budgetary or financial implications.

The only way out of the impasse is for a strong core of countries, led by France and Germany, to finally put concrete proposals on the table, enabling progress to be made on both budgetary and institutional fronts, without waiting for unanimous agreement from the other countries. The concept of a strong core to overcome the obstacles of unanimous consensus has been raised many times in the past, most recently in the report by Mario Draghi proposing a massive investment plan for Europe.

The time has now come to give it substance and make real progress. For this to happen, three conditions have to be met: This core group must be given solid institutional and democratic foundations; it is essential that Germany, and not just France, Italy, or Spain, should be able to play its part, particularly on the budgetary front; and within each country, and at the European level as a whole, several political visions, from both the right and left, must be able to express themselves and flourish.

Let's start with the first point. To create a core group of countries capable of making important budgetary and financial decisions with all the necessary democratic legitimacy, it is important to base it on a solid institutional and political framework. The most logical starting point would be the Franco-German Parliamentary Assembly (FGPA), set up in 2019 as part of the renewal of the bilateral Franco-German treaty. A young and little-known institution, made up of 100 lawmakers from all the parliamentary groups in the Assemblée Nationale and Bundestag, the FGPA has met between two and three times a year since its creation and has so far been confined to a mainly consultative role. But there's nothing to prevent the two countries from giving it an essential decision-making role, particularly on budgetary matters, and opening it up to all EU countries wishing to join this strong core, thus transforming it into a genuine European assembly, as in the draft "Manifesto for the Democratization of Europe" (tdem.eu).

This strengthened parliamentary union, which could be called a "European parliamentary union" (EPU), would bring together within the EU those countries ready to unite further to influence world affairs and invest in the future, and in particular to borrow jointly to finance investments in energy, transport, research, and new technologies.

Let's turn to the second point. While some countries, such as France, Italy, and Spain, can identify with such a vision, the central difficulty has always been to convince Germany, which is very reticent about the idea of joint borrowing, or even borrowing in general. This situation is now changing: A growing segment of the German public understands that the country urgently needs to invest in its infrastructure, in the disadvantaged regions of the East as well as throughout the country. This is now the majority opinion of German economists, recently joined by a large proportion of employers. In fact, the issue is in the process of shattering the coalition between the left and the liberals.

To overcome the last remaining reservations, however, it's necessary to demonstrate that joint European borrowing is the right tool and guarantee that it will not lead to a "transfer union," an absolute bogeyman on the other side of the Rhine. For example, it can be stipulated in advance that transfers will not exceed 0.1% of GDP (article 9 of the draft manifesto; see tdem.eu).

This is unfortunate: If students from all over Europe are going to attend the same campuses to prepare for the future of the continent, then one day we'll have to stop complex calculations and address all Europeans indiscriminately. But in the meantime, we need to find ways to move forward and build trust between countries and states with different histories. It's worth the effort. The euro and the European Central Bank (ECB) now have such a global financial footprint that it's in everyone's interest to agree to borrow together, even without transfers.

Third and last point: For a project of such importance to see the light of day, it is essential that several political visions can be brought together. In the Draghi report, the approach is deliberately liberal and technophile. The former ECB president insists on public subsidies for private investment, for example in industry or artificial intelligence, and on clusters of excellence in major metropolises. Liberals and the right will also be in agreement and will no doubt emphasize the strengthening of military budgets and the promotion of a "Fortress Europe."

The left, on the other hand, will place the emphasis on social, educational, and health investments, and infrastructure open to the greatest number of people, in poor suburbs as well as in outlying regions, as well as on tax justice objectives. It will be up to the European assembly and national parliaments to decide, on the basis of contradictory deliberations, under the watchful eye of Europe's citizens.

What does it matter if this reinforced union begins with just a few countries? The urgent need today is to react to the shock of Trumpism by asserting the strength of European values—those of parliamentary democracy, the social state, and investment in the future.

For a New Left-Right Cleavage
December 17, 2024

Let's face it: France will not emerge from its current political crisis by inventing a new central coalition. The idea that the country should be governed by bringing together all the so-called reasonable parties, from the center-left to the center-right, from the Parti Socialiste (PS) to Les Républicains (LR), excluding the "extremes"—La France Insoumise (LFI) on the left and the Rassemblement National (RN) on the right—is a dangerous illusion, which will only lead to further disappointment and strengthen the extremes in question. Firstly, because this coalition of the reasonable looks very much like a coalition of the better-off. Excluding the working class from government is certainly not the way out of the democratic crisis. Secondly, electoral democracy needs clear and accepted alternations to function properly.

The virtue of the left-right bipolarization, provided that its content is sufficiently rapidly renewed in the face of global change, is that it makes such alternations possible. Two coalitions driven by antagonistic but coherent visions of the future and based on divergent socio-economic interests alternate in power, and this is how voters can form their opinions, adjust their votes, and have confidence in the democratic system itself. It was this virtuous model that enabled the consolidation of democracy throughout the twentieth century, and it is toward a new left-right bipolarization that we must move today if we wish to avoid democratic disintegration.

Once that's acknowledged, how do we proceed? In countries with a first-past-the-post system, such as the United States and the United Kingdom, bipolarization is a matter of course. But you have to know what kind of bipolarization you want. Across the Channel, Labour has replaced the Conservatives, but with a program so timid that it is already generating mistrust, having won power with a low score and thanks to very strong divisions on the right. Across the Atlantic, bipolarization has turned in on itself. After abandoning any redistributive ambitions, the Democrats have become, over the last few decades, the party of the most highly educated, and of the highest income earners.

The Republicans retain a strong support base among the business world, but they have succeeded at little cost in attracting the popular vote by breaking with the Democrats' dogma of free trade and liberal, urban, and elitist globalization. Time will tell whether a new turnaround is possible. What is certain is that it will require a major change of course for the Democrats. In a democracy, it's impossible to obtain your best scores among both the most privileged and the most disadvantaged. If the Democrats want to become the party of social justice once again, it must accept the loss of the vote of the privileged by proposing vigorous redistribution measures, which will have to respond not only to the aspirations of the urban working class but also to that of small towns and rural areas. For example, you can't bet everything on canceling student debt: You also need to reach out to those who have taken on debt to buy a home or a small business.

In the French context, these kinds of questions are posed differently. The left is still very much alive, but here too it has lost the popular vote in the towns and villages hard hit by trade liberalization, deindustrialization, and the absence of public services. In 1981, the left scored virtually the same among the working class, whatever the size of the town or city. Over the last few decades, a territorial divide on a scale unseen for a century has emerged between the urban working class (who continue to vote left) and the rural working class (who have switched to the RN). It is on this territorial divide that the political tripartition was built: An objectively very privileged central bloc intends to govern the country based on the divisions between the urban and rural working classes between the left and right blocs.[1]

The presidential election can play a useful role in breaking out of this tripartition: It encourages people to come together in the second round, and can speed up the advent of a new bipolarization. But that won't be enough: What we need above all is in-depth work within the political parties and the citizens who support them. On the left, the parties must learn to deliberate and settle their differences democratically, first by deputies from the Nouveau Front Populaire left-wing coalition voting among themselves, then by directly involving left-wing voters. The priority must be to respond to the aspirations of the working classes in all territories, and to rally people far beyond each party's electoral base. In particular, LFI will have to show humility and accept a simple fact: Its current electoral core is real, but it is a minority in the country, and can hardly envisage victory in the second round.

On the right, it's time for LR and the more right-wing factions of the Macronist camp to accept the idea that they need

---

1 J. Cagé and T. Piketty, *A History of Political Conflict: Elections and Social Inequalities in France, 1789–2022* (Belknap Press of Harvard University Press, 2025).

to form a majority coalition with the RN. This is already what they did when they voted for the immigration law and many other texts (such as the anti-tenant law). It's time to openly embrace the union of the right, otherwise it will sooner or later be imposed at the ballot box. This is also what will force the RN to move away from easy posturing and to shift its economic and budgetary discourse to the right, contributing to the emergence of a new bipolarization. What is certain is that it is not healthy to leave LFI and RN outside the system: They must both assume their place within their respective coalitions and face the difficult test of power. This is the price democracy will have to pay to emerge from its current crisis.

Democracy vs. Oligarchy, the Fight of the Century
January 21, 2025

A few days ahead of Donald Trump, Elon Musk, and tech executives aligned with the Make America Great Again (MAGA) movement coming to power, Joe Biden delivered a forceful warning about the emergence of a new "tech industrial complex" threatening the United States' democratic ideal. For the outgoing president, the extreme concentration of wealth and power risked undermining "our entire democracy, our basic rights and freedoms, and a fair shot for everyone to get ahead."

Biden is not wrong. The issue is that he has done little to oppose the oligarchic drift taking place both in his country and globally. In the 1930s, his predecessor Roosevelt, also deeply concerned about such trends, did not stop at making speeches. Under his leadership, Democrats implemented a robust policy of reducing social inequalities (with tax rates on the highest incomes nearing 70%–80% for half a century) and investing in public infrastructure, health, and education.

In the 1980s, Republican Ronald Reagan, deftly playing on nationalism and a feeling of catching up, undertook to dismantle Roosevelt's New Deal. The problem was that Democrats, far from defending this legacy, actually helped legitimize and solidify Reagan's turn, notably under the Clinton (1993–2001) and Obama (2009–2017) administrations.

Biden has often been described as more of an interventionist than his predecessors in economic matters. This is not entirely false, minus two major drawbacks. Biden was among the Democrats who voted for the Tax Reform Act of 1986, the foundational law of Reaganism, which dismantled Roosevelt's progressive tax system by lowering the top tax rate to 28%. Everyone can make mistakes, but Biden has never felt it necessary to explain that he had made a mistake or changed his mind. If spending isn't funded, inflation inevitably rises, another major issue on which we are still awaiting Biden's remorse.

Moreover, the outgoing administration's so-called Inflation Reduction Act primarily facilitated the flow of public funds into private enterprises, effectively supporting the accumulation of private capital. There is no doubt that the Trump administration will push this unrestrained alliance between the federal government and private interests to its peak.

Could Democrats change course in the future? The overwhelming influence of private money in US politics, as pervasive among Democrats as it is among Republicans (if not more so, even with the recent growth of small donations), urges caution. However, the party's chances of finding its footing remain real. First, the mix of nationalism and ultra-liberalism taking power in Washington will solve none of the social and environmental challenges of our time. Second, opposition to oligarchy continues to be a cornerstone of the nation's identity.

In 2020, the Bernie Sanders–Elizabeth Warren duo had proposed extending Roosevelt's New Deal, with the addition of a mega-wealth tax (with rates reaching 8% annually on billionaires, a level never seen in Europe), a massive investment

Democracy vs. Oligarchy, the Fight of the Century

plan for universities and public infrastructure, and the invention of a US-tailored economic democracy (with significant voting rights for employees in corporate boards, as practiced in Germany and Sweden for decades). The two candidates had nearly tied with Biden and won overwhelmingly among younger voters. Disillusioned by the Biden-Harris experience, Democrat voters were largely absent in 2024, a costly blow to the party. It is entirely possible that a Sanders–Warren-style candidacy could succeed in the future.

Above all, the rest of the world might well spearhead the most progressive political changes in the decades to come. Little is expected from the authoritarian oligarchies that China and Russia have become. But within the BRICS, there are vibrant democracies representing more voters than all Western countries combined, starting with India, Brazil, and South Africa. In 2024, Brazil supported the idea of a global wealth tax on billionaires at the G20.

The initiative was unfortunately rejected by the West, who, in the same year, also found it strategic to oppose a proposed UN tax convention, in an effort to maintain their monopoly over international tax cooperation within the rich-country club of the OECD. This stance also sought to avoid any significant redistribution of revenue on a global scale. If, in the coming years, India shifted to the left and sent the nationalist, business-oriented BJP into opposition—an increasingly plausible scenario—the pressure from the Global South for fiscal and climate justice could become irresistible.

In the global battle between democracy and oligarchy, one can only hope that Europeans will emerge from their lethargy and play their full role. Europe invented the welfare state and the social-democratic revolution in the twentieth century, and it stands to lose the most from Trumpist hypercapitalism. Here again, there is reason for optimism: Since the

Covid-19 pandemic, the public expects a lot from the European Union and is less hesitant than its leaders. One can only hope that these leaders will rise to the occasion and, by 2025, manage to overcome the mutual distrust and perpetual self-criticism that has held them back.

Index

Locators in *italics* indicate figures and tables.

Abdelal, Rawi, 180
Abu Ghraib prison, abuses at, 75
AEA (American Economic Association), 150, 151
AFD (Alternative for Germany) (right-wing party), 41
Afghanistan, 76, 78
Africa, 43, 60, 87, *88*, 101
agribusiness, 127
agricultural crisis, 183
Amsterdam, Treaty of, 163
anti-Semitism, 197
Argentina, 172
Arthaud, Nathalie, *33, 34*
artificial intelligence, 209
Asia, 43, 101
Asian financial crisis (1997), 181
Assemblée Nationale (France), 23, 41, 203, 216
austerity, fiscal, 208

Babis, Andrej, 81
Balladur, Édouard, 163
Barnier, Michel, 213–214
basic income, 62, 63, 64

Basic Law (Germany, 1919), 187
Bauluz, Luis, 87
Belle Epoque (1870–1914), 202
Benhenda, Asma, 57
Besancenot, Olivier, *33, 34*
Biden, Joe, 38, 59, 223, 224, 225
billionaires, 2, 20, 77, 145, 204; debates over taxation of, 211–214; evasion of taxes by, 79; media owned by, 116; structural under-taxation of, 181; tax rate in United States, 128, 129; wealth tax on, 38, 51, 60, 209, 225
biodiversity protection, 12
Bitcoin, 63
BJP (Bharatiya Janata Party) (India), 44, 225
Blair, Tony, 39, 81
Blanchet, Thomas, 87
Bolloré, Vincent, 205
Bolsheviks/Bolshevism, 22, 25, 26, 123
Borloo report (2018), 161
Borne, Elizabeth, 136

Bosnian federation, 169
Brazil, 44–45, 117, 199, 225
Bretton Woods institutions, 172
Brexit referendum (2016), 109, 119
BRICS (Brazil, Russia, India, China, South Africa) group, 45, 171–174, 199, 225
Brief History of Equality, A (Piketty), 1
Britain. *See* United Kingdom
BSW (Bündnis Sahra Wagenknecht–Vernunft und Gerechtigkeit) (German political party), 41–42
building renovations, energy-efficient, 12, 49, 64, 129, 155, 201, 208

CAF (Caisse d'Allocations Familiales), 175–177
Cagé, Julia, 32, 162
Canada, in G7 group, 171
capital, free circulation of, 37, 38, 60, 67; economic illusion of, 77; sacralization of, 131; taxation and, 81, 212; as unsustainable system, 98
Capital au XXIe Siècle, Le (Capital in the twenty-first century) (Piketty), 45
capitalism, 2, 4, 13, 47, 107; agrarian and industrial, 120; challenge from China to, 70; crises of, 26; ecological catastrophe and, 28; hyper-capitalist ideology, 97, 225; overcoming of, 12, 16, 27, 38
Capital Rules (Abdelal), 180
carbon emissions, 2, 72, 90; in China, 201; customs duties and, 131; EPU and taxes on, 156; reduction in, 4; social class and, 3; taxes on, 126
carbon footprint, 84
carbon tax, 29, 84, 92, 93, 134
CCP (Chinese Communist Party), 70
CDU (Christian Democratic Union) (Germany), 23, 41, 51, 124; Kirchoff affair and, 147; liberalization of capital flows advocated by, 180
CFDT (French trade union), 49
Challenges magazine, 51, 79, 107
Chancel, Lucas, 90
Charles I, beheading of (1649), 120
China, 4, 8, 58, 132; as authoritarian oligarchy, 225; in BRICS group, 45, 199; Chinese oligarchs, 95, 97; decline of public property in, 71; destruction of electoral pluralism in Hong Kong, 117, 131, 201; Global South and, 117; Olympic Games (2008) in, 130; ownership of firms in, 73; as part of Global North, 145; as perfect digital dictatorship, 71, 172; rising tensions with the West, 115; weaknesses and assets of, 72–73; women's share in labor income, 90
Chirac, Jacques, 163
Christians, evangelical, 197
Citizens United decision (2010), 147
civilizations, clash of, 75, 76, 78
Civil War, US, 105
climate change/crisis, 2, 4, 84, 123; inequalities and, 90; responsibility for, 3; tax rate on high

Index

incomes and, 129. *See also* global warming
Clinton, Bill, 38, 39, 224
colonialism, 72
communist parties, 24, 44. *See also* Parti communiste français
Congress Party (India), 44
Conseil de la Résistance, 23
Constitution, of Weimar Republic, 187–190
constitutional judges, abuse of power and, 146–149
construction sector, 5, 10
consumption, climate crisis and, 2
Corbyn, Jeremy, 38–39
Covid pandemic (2020–2021), 26, 62, 87, 124, 156, 226; EU recovery plan for, 207; European Central Bank (ECB) and, 180; international solidarity and, 58; social justice and, 49
Crimea, 99, 200
CRS (Common Reporting Standards), 52
CSG (generalized social contribution) (France), 17n11, 141, 143

Debré, Jean-Louis, 147
debt, public, 50, 51, 73–74, 188, 208, 211
decommodification, egalitarian, 2–5, 24, 25, 64; extension into new sectors, 37; march toward equality and, 5–9
Defender of Rights, 54, 56
De Gaulle, Charles, 163
deindustrialization, 181, 221

Delors, Jacques, 179, 180
democracy, electoral/parliamentary, 2, 21, 25, 115, 155, 218; "autocracy" as adversary of, 97, 98; electoral abstention by poorest population, 117; limits of, 124; referendum democracy, 85–86; socialism as pillar of, 45–47; in struggle against oligarchy, 223–226; three-tier democracy, 111–114; Ukraine and, 191
Democratic Party (United States), 14, 38, 39, 105, 220, 223–225
democratic socialism, 35, 36, 39
Denmark, 17n11, 62
deregulation, 152, 159
development, equal minimum rights to, 60, 77, 118
development, sustainable, 4, 183
"dictatorship of the proletariat," 25
digital currencies, 63
digital technology, 209
discrimination, fight against, 84
Donbas region (Ukraine), 99, 200
Draghi, Mario, report of, 207–210, 215, 217–218
Dumont, René, *30, 31*
dumping, 123, 193; climate, 132; environmental, 109, 125, 131, 182; fiscal, 109, 125, 182; tax, 40, 132, 181, 193
Dupont-Aignan, Nicolas, 111

EA (European Assembly), 156
ecological (green) politics: failure of ecologism without socialism, 28–35; territorial divide in elections and, 30–31, *30*

Ecologists, French, 85, 91, 106
Economic Bill of Rights (United States, 1944), 64
economics, 152–153
education sector, 10, 16, 124, 128; decommodification of, 47; democratization of, 11; equal minimum rights to, 60, 77, 118; in France, 23–24; freezing of public resources for, 27; inequalities in, 57; investment in, 74; spending on, 5, 6
EEC (European Economic Community), 179
Egypt, 45, 57, 172, 199
elections, territorial divide and, 30–33, *30*
electric vehicles, 131
Elizabeth II, queen, 119
energy sector, 5, 10, 16, 27; capitalism and, 7; decommodification of, 47; EPU and, 155; investment in, 208; renewable energy, 12, 49
Ensemble (French liberal political coalition), 163, *164*
environmental crises, 3, 26, 37, 131, 144–145
environmental protection, 10, 27
EPC (European Political Community), 125, 154, 155
EPU (European Parliamentary Union), 125, 154–157, 182, 216
Ethiopia, 45, 172, 199
EU (European Union), 50, 122, 124, 126, 131, 182, 226; Carbon Border Adjustment Mechanism, 193–194; carbon tax of, 201; decision-making by unanimous vote of member states, 215; EPC (European Political Community) and, 154, 155; EPU (European Parliamentary Union) and, 156; founding of, 179; investment plan for the future, 207–210, 215; Israeli-Palestinian conflict and, 169–170, 195, 196; Lisbon Treaty and, 132; military budget of, 199; peace and, 169; Ukraine as possible member of, 191–194
European Central Bank (ECB), 85, 124, 180, 181, 207, 217–218
European Court of Justice (ECJ), 67
European Fiscal Compact (2012), 180
European Tax Observatory, 67, 84, 109–110, 150, 193
eurozone, 155, 181, 210

farmers: average annual income of, 183, 184; FNSEA farmers' union, 185; as self-employed people, 184
farming, organic, 10, 186
Fascism, 123
FDP (Free Democratic Party) (Germany), 41
federalism, 123–126, 133
FGPA (Franco-German Parliamentary Assembly), 155, 216
financial capital, 3
financial crisis (1929), 26
financial crisis (2008), 26, 37, 39, 124; European Central Bank (ECB) and, 180; IMF and, 181
Fisher, Irving, 151
Floyd, George, 54
France, 40, 57, 100, 193, 210, 215; composition of property in, 63, 65;

decline of public property in, 71; ecological movement in, 29; education in, 144; English revolutions compared with, 119–122; EPU and, 155; Front Populaire in, 23; in G7 group, 171; peace with Germany in the EU, 169; presidential election, 81–86, 103–104; *taxe foncière* (tax on asset ownership), 80–81; tax on multinationals, 67; tax rate on high incomes, 18, 128, 129; territorial divides in, 156–161; three-tier democracy in, 111–114, 165, 166; wealth disparity in, 128, 134, 211; welfare state in, 124
Freiburg school, 123
French Revolution, 80
Friedman, Milton, 75, 152
Front Populaire (Popular Front) (France), 23

gas engines, banning of, 5, 16
Gaullists, 23, 108
Gaza Strip, 167, 168, 169, 171, 199
GDP (gross domestic product), 64, 100, 101, 210; of BRICS countries, 171; global tax on high incomes and, 60; health spending in United States and, 204; property tax and, 80; of richest and poorest *départements* of France, 158–159; top French fortunes' share of, 51, 79, 107–108, 128; "transfer union" and, 217
geopolitics, 47, 199–202, 215
Germany, 12, 40, 57, 91, 210, 215; decline of public property in, 71; disadvantaged regions in, 209; ecological movement in, 29; EPU and, 155; former East Germany, 41; in G7 group, 171; Kirchoff affair in, 147; *Lastenaugleich* ("burden-sharing") system, 51, 188; peace with France in the EU, 169; tax rate on high incomes, 18; Weimar Republic, 187–188; welfare state in, 124
GFR (global financial registry), 97
Giscard d'Estaing, Valéry, 163
Gladstone, William, 120
Global Inequality Report 2022, 128
globalization, 47, 93, 212, 220
Global North, 2, 68, 129; neoliberalism and, 74; revenue sharing and, 77; transparency on excessive enrichment in, 77
Global South, 2, 68, 69, 129; incomes of working poor in, 62; Israeli-Palestinian binational state and, 196; neoliberalism and, 74; pressure for fiscal and climate justice from, 225; revenue sharing and, 77, 84; right to development and self-governance, 78; taxation in, 80–81; transparency on excessive enrichment in, 77
global warming, 9, 84, 117, 138. *See also* climate change/crisis
"Glorious Revolution" (Britain, 1688), 120
Great Transformation, The (Polanyi), 8n5
Greece, 40
Green New Deal, 64
green parties, 29–30, 33, 35, 41
Green Party (Germany), 41, 42

G7 countries, 66–69, 171
G20 countries, 44, 45, 211, 225

Halde (Haute Autorité de lutte contre les discriminations et pour l'égalité), 54
Hamas, 167, 196, 197
Harris, Kamala, 38, 225
Hayek, Friedrich, 22, 123, 152
healthcare, 10, 16, 124, 128; decommodification of, 47; EPU and, 155; equal minimum rights to, 60, 77, 118; in France, 24; freezing of public resources for, 27; growth of public resources in, 11; investment in, 208; spending on, 5, 6; in United States, 8; universal health coverge, 46
Henry VIII, king, 119
Hindenburg, Paul von, 189
History of Political Conflict, A (Cagé and Piketty), 162
Hohenzollern princes, expropriation of, 187, 189
Hong Kong, 71, 97, 117, 131, 201
hospitals, 7, 11
House of Commons (Britain), 119, 121
House of Lords (Britain), 119, 120, 121
housing, 7, 16, 27, 128, 160; decommodification of, 47; renovation of, 134; right-wing anti-poor ideology and, 177
Hughes, Michael, 51
human capital, 7, 81
human rights, 77
Hungary, 105
Huntington, Samuel, 75
hydrocarbons, 4, 100

identity issues/conflicts, 1, 28, 54, 104; as distraction from socio-economic issues, 178; fall of communism and, 113; nationalist politics and, 160
IMF (International Monetary Fund), 44, 172, 181
immigration, as political issue, 41, 42, 93, 103, 104; anti-migrant right, 127; white working classes and, 105
income gaps, 7, 76
India, 4, 26, 117; basic income proposal in, 62; in BRICS group, 45, 199, 225; caste system in, 57; electoral democracy in, 172; income and discrimination in, 55; vaccine production capacity in, 58; as world's most populous country, 138
Indonesia, 45
Industrial Revolution, 88, 112
inequalities: gendered wage, 136; pay within self-employed occupations, 185; reduction in, 2, 4, 28, 32; rise in, 72; territorial divide and, 159–160
inflation, 49–50, 51, 101, 107, 181, 189
inheritance, 65, 89, 213
Insoumis, 85, 91, 106, 108. *See also* LFI (La France Insoumise)
interest rates, 49, 208
internationalism, 42, 78, 94, 106, 182
Iran, 45, 172, 199
Iraq, US invasion of (2003), 75–76, 116, 171
Ireland, 67, 120
ISF (*impôt sur la fortune*) (wealth tax), 92, 108, 128, 204

Index

Islamic State, 76
isolationism, 77, 78
Israeli-Palestinian conflict, 167–170, 171; binational state as solution, 195–198; geopolitics and, 199. *See also* Gaza Strip
Italy, 40, 57, 127, 210, 216; EPU and, 155; in G7 group, 171

Jadot, Yannick, *30, 31*
Japan, 8; decline of public property in, *71*; in G7 group, 171; tax rate on high incomes, *18*
Joly, Eva, *30, 31*

Kaneworthy, L., 36n19
King, Martin Luther, Jr., 64
King, Willford, 151
Kirchoff, Paul, 147
KPD (German Communist Party), 189

labor, international division of, 77
Labour Party, British, 15, 28, 38–39, 92, 220
Laguiller, Arlette, *33, 34*
Lalonde, Brice, *30, 31*
Lamy, Pascal, 180
"Land for All, A," coalition, 167–168, 169, 196
landlords, 12
Lassalle, Jean, 111
Latin America, 26, 43, 87, *88*
LCR (Ligue communiste révolutionnaire) (French Trotskyist group), *33, 34*
left-right bipolarization, 165, 166, 219–222
left-right political divide, 103–106
left-wing parties, 35, 39, 111, 161, 203; alliance with green parties, 41; French presidential election and, 91–92; in India, 44; "New Popular Union" and, 107–110; support in urban areas for, 32; young people of immigrant background as supporters, 113–114
Lenin, Vladimir, 25
Lepage, Corinne, *30, 31*
Le Pen, Jean-Marie, 178
Le Pen, Marine, 103, 105, 111
LePenism, 92
LFI (La France Insoumise), 32, 40, 92, 206, 219, 221, 222. See also *Insoumis*
liberalism, 2, 37–38, 43, 178; center-right bloc in France, 111, 112; coalition of the left and liberals, 217; fall of communism and, 113; as part of three-pillar democracy, 45–47
Liberals, British, 120, 121
life expectancy, 136
Lisbon Treaty (2007), 132, 179
Lloyd George, David, 120, 121
LO (Swiss trade union confederation), 13
LO (Lutte Ouvrière) (French Trotskyist group), *33, 34*
LR (Les Républicains) (French right-wing party), 93, 136, 177, 219
Luxembourg, 49, 50, 67, 117
LuxLeaks (2014), 79

Maastricht Treaty (1992), 124, 156, 179, 180, 192
Macron, Emmanuel, 92, 93, 103, 108, 176; bourgeois votes for, 163; centrism of, 104; liberal/center-right bloc and, 111;

Macron, Emmanuel (*continued*)
    pension issue and, 104–105,
    142–143; as president of the rich,
    134; raising of retirement age
    and, 205; tax justice issue and,
    105
Macronism, 83, 91, 92, 93, 104
MAGA (Make America Great
    Again) movement, 223
Mamère, Noël, *30, 31*
Manifesto for the Democratization of Europe, 125, 156,
    216
March for Jobs and Freedom
    (United States, 1963), 64
markets, sacralization of, 8n5, 42
Marshall Plan, 207
Martinez-Toledano, Clara, 87
Marx, Karl, 25
McCutcheon decision (2014), 147
media, 38, 79, 205; democratization of, 21; exposé of tax havens,
    49; financing of, 20, 156; ownership by billionaires, 116; in
    Ukraine, 191, 192
Meidner, Rudolf, 13
Mélenchon, Jean-Luc, 103
Merkel, Angela, 147
M5S (Five Star Movement), 40
Micheletti, Pierre, 60
middle classes, 2, 17, 108; climate
    challenge and, 134; "lower
    middle classes," 47; progressive
    taxation and, 19; social-
    ecological program and, 127
Middle East, 4, 87, *88*
minimum wage, 64, 140, 147, 185
Ministry for the Future, The
    (Robinson), 173–174
Mitterrand, François, 180
multi-millionaires, 96, 98, 173, 181,
    209

multinational corporations, 59, 67,
    117, 145, 204; EPU and taxes on,
    156; minimum tax on, 59–60,
    84, 173; reform of taxation on, 77
municipal companies, 7
Muslims, political sentiment
    against, 93, 103, 104, 197

National Front–National Rally
    (FN-RN) (French far right
    party), 160, 163
National Health Service (Britain),
    121
nationalism, 2, 43, 93, 224;
    nationalist (extreme right) bloc
    in France, 111, 112; as part of
    three-pillar democracy, 45–47;
    reified ethno-national solidarities and, 112; right-wing, 40;
    xenophobia and, 94
nation-states, 4
NATO (North Atlantic Treaty
    Organization), 100, 115–116,
    200
natural resources, exploitation of,
    8–9
Nazism, 123, 190
Neef, Theresa, 89
neo-conservatives, US, 75
neoliberalism, 1, 28, 46–47, 74
Netanyahu, Benjamin, 195
New Deal, 146, 152, 223
NFP (Nouveau Front Populaire)
    (France), 41, 203–205, 221
NHS (National Health Service)
    (Britain), 22
NPA (Nouveau parti anticapitaliste) (French Trotskyist
    group), *33, 34*
NUPES (Nouvelle Union
    Populaire, Écologique et Social)
    (France), 41

Obama, Barack, 38, 224
Observatory of Discrimination, 54, 55, 56
Ocasio-Cortez, Alexandria, 14
OECD (Organisation for Economic Co-operation and Development), 50, 52, 59, 60, 150, 193, 225
oligarchs, Russian and Chinese, 95, 97–98, 101, 117, 138
OpenLux survey, 50

Palestine. *See* Gaza Strip; Israeli-Palestinian conflict; West Bank
Pandora Papers, 79, 81
Paradise Papers (2017), 79
Parcoursup, 205
Parti communiste français (PCF) (French Communist Party), 25n14, 32, 85, 91, 106
Pasok (Greek socialist party), 40
Pécresse, Valérie, 93, 111
pension reform, 138–141, 142, 143, 148
pensions, 8n6, 135–137, 205
People's Budget crisis (United Kingdom, 1909), 119, 120
pesticides, 183, 185
petro-monarchies, 78, 117
Pinochet, Augusto, 22
plastic, banning of, 5, 16
Plessy v. Ferguson (1896), 146
Podemos (citizen coalition in Spain), 40, 92
Poland, 105, 166
Polanyi, Karl, 8
pollution, 183
poor, stigmatization of, 177
Poutou, Philippe, *33, 34*
power relations, 1, 48
power sharing, *14*, 74, 84, 112
production, means of, 24, 188

productivity, 7, 144, 204
property, concentration of, *63*, 65
property rights, 188
ProPublica survey, 66, 68, 79, 80, 128
prosperity, economic, 8, 13, 124; access to human capital and, 7; education and, 129, 144; factors involved in, 17–18
protectionism, 130–133, 193
PS (Parti socialiste) (French Socialist Party), 28, 32, 85, 91, 92, 106
PSOE (Partido Socialista Obrero España) (Spanish socialist party), 40
public opinion, 138, 183, 194
public service, 5, 175–178
public spending, freezing of, 27, 36
Putin, Vladimir, 95, 96, 101

Qatar, World Cup in, 130

racism, 54–57, 114
Radicals, in France, 32
Rawls, John, 152
Reagan, Ronald, 223
real estate capital, 3
reconstruction, post-war, 51
redistribution, 21, 22, 27, 88–89; abandoned by US Democratic Party, 220; of economic power, 205; environmental challenges and, 48; G7 countries and, 68; of inheritance, 65, 89; as planet-saving process, 127–129; in platform of French left-wing parties, 32; proposals for, 62
Renaissance (Macron's party), 177
Republican Party (United States), 105, 220, 224
retirement age, 139, 205

RN (Rassemblement National) (French right-wing party), 136, 165, 177–178, 181, 219, 221–222
Road to Serfdom, The (Hayek), 22
Robillard, Anne-Sophie, 89
Robinson, Kim Stanley, 173–174
Rome, Treaty of (1957), 179
Roosevelt, Franklin D., 64, 146, 151, 223, 224
RPR-UDF (French conservative political coalition), 163
rule of law, 116, 148
Russia, 4, 58; as authoritarian oligarchy, 225; in BRICS group, 45, 199; NATO's strength measured against, 200; as part of Global North, 145; public and private assets in Europe, 202; sanctions against, 95–98, 138; World Cup (2018) in, 130. *See also* Ukraine, Russian invasion of

Salisbury, Lord, 120
Sanders, Bernie, 14, 38, 152, 224
Sarkozy, Nicolas, 163, 176
Saudi Arabia, 78, 172
schools, 7, 11
September 11, 2001, terrorist attacks, 75–78
"shell" companies, 50, 81
Silva, Lula da, 44, 127
Single European Act (1986), 179, 192
slavery, 72, 116
SMEs (small and medium-sized enterprises), 109
social democracy, 1, 5, 48; toward democratic and ecological socialism, 35–37; lost revolutionary momentum of, 26–28; as revolutionary movement in the twentieth century, 21–26; "social-democratic revolution," 6, 9, 15, 26. *See also* SPD (Social Democratic Party) (Germany)
Social Democratic Party, Brazilian, 36n20
Social Democratic Party, Portuguese, 36n20
Social Democratic Party, Russian, 25
Social Democrats, Swedish, 22
Social Democrats, Swiss, 35
socialism, democratic and ecological, 1, 2, 35, 42; Global South and, 43–44; gradual decommodification and, 47
Socialist Party (Portugal), 36n20
Socialist Party, French. *See* PS (Parti socialiste)
social justice, 22, 51, 117, 200, 201; abandonment of ambition for, 82, 84; Covid pandemic and, 49; rebuilding of the left and, 204; US Democratic Party and, 220
social security, 5, 7, 11, 23, 184; neoliberalism and, 46; racialized recipients of, 47; Sarkozy's hunt for fraud in, 176
sociology, 152
South Africa, 45, 117; in BRICS group, 45, 199, 225; income and discrimination in, 55; vaccine production capacity in, 58
sovereignty, 123, 132, 212; fiscal, 82, 84; Palestinian, 168; universalist, 78
Soviet Union, 8, 37
Spain, 91, 155, 210, 216
SPD (Social Democratic Party) (Germany), 23, 28, 42; coalition

Index

with other parties, 41, 147;
  expropriation of Hohenzollerns
  and, 189; years in power, 92
Starmer, Keir, 39
Summers, Larry, 152
Sunak, Rishi, 39
Supreme Court, US, 116, 146–147
Sweden, 12, 22, 57, 108, 127
Swift financial network, 96
Switzerland, 117, 154, 213
Syriza (Greek left-wing coalition),
  40

Taiwan, 71
Taliban, 76, 78
tariffs, 131, 132
taxation, 5, 59; abolition of income
  tax, 178; abolition of the wealth
  tax, 134; carbon tax, 29, 84, 92,
  93, 134; climate change/crisis
  and, 84; direct, 5, 17; evasion
  of, 29, 50; free movement of
  capital and, 81, 212; French
  constitution and, 147–148;
  indirect, 5; property tax, 89,
  148; public statistics on, 52;
  registration of property and,
  80; regressive, 29, 81, 136; share
  of national income, 8n6, 24, 27;
  taxe foncière (tax on asset
  ownership), 80–81; tax justice,
  52, 105, 139, 211, 218; wealth tax
  proposals in United States, 152;
  white-collar tax fraud and
  evasion, 176. *See also* ISF (*impôt
  sur la fortune*)
taxation, progressive, 9, 16, 17,
  19, 89; basic income and, 63;
  global minimum wealth tax,
  44–45; G7 countries and, 68;
  inheritance and, 108; minimum

tax on multinationals, 59–60; as
  percentage of GDP, 60; of
  Roosevelt administration, 224;
  socialization of wealth and, 20,
  21; tax compliance and, 19;
  undermining of, 20; Victory
  Tax (United States), 151
tax havens, 49, 59, 66, 67, 93, 110,
  181
Tax Reform Act (United States,
  1986), 224
Tcherneva, Pavlina, 64
television channels, 20
"Theory of Justice" (Rawls), 152
think tanks, 20
Thorez, Maurice, 25n14
Tories/Conservatives, British, 39,
  120, 220
trade unions, 20, 22; Social
  Democratic parties and, 23, 25;
  trade union rights, 65, 130, 188
transportation, 5, 6, 10, 16, 128;
  capitalism and, 7; decommodification of, 47; EPU and, 155;
  investment in, 208; local public
  transportation, 12; in small towns
  and rural areas, 27
Treaty establishing a Constitution
  for Europe (2005), 179
Trente Glorieuse ("Glorious
  Thirty") (period of prosperity
  in France, 1950–1980), 72, 89
Trotskyists, 32, 33, *33*, 34
Trump, Donald, 215, 223
Trumpism, 74, 173, 218, 225
Truss, Liz, 122
Turkey, 56

UDI (L'Union des démocrates et
  indépendants), *164*
Ukraine, EU and, 191–194

Ukraine, Russian invasion of, 95–98, 138; geopolitics and, 199; historical analogies to, 99–100; NATO intervention as possibility, 200; Western democratic model and, 115–116; Western response to, 100–102
UN (United Nations), 44, 101, 138, 168, 225
unemployment benefits, 8n6, 65
Union de la Gauche (France), 23
United Arab Emirates, 45, 172, 199
United Kingdom (Britain), 22, 42, 91, 100; decline of public property in, *71*; education in, 144; ethno-racial categories in, 56, 57; first-past-the-post system in, 220; in G7 group, 171; social and democratic revolution in, 119–122; tax rate on high incomes, *18*
United States, 7–8, 26, 42, 115; China perceived as threat, 78; decline of public property in, *71*; ethno-racial categories in, 56, 57; first-past-the-post system in, 220; fragile democracy in, 173; in G7 group, 171; income and discrimination in, *55*; Israeli-Palestinian conflict and, 195, 196; progressive taxation in, *19*; sanctions against Russia, 96; Swiss bank licenses and, 212; tax rate on high incomes, 17–18, *18*; "tech industrial complex" and democracy in, 223; wealth disparity in, 88; as world's educational leader, 144
universities, 7, 11, 83–84, 201; financing of, 209; investment in public universities, 38; "Islamo-leftism" in, 93; Israeli-Palestinian conflict and, 195; US Democratic Party investment plan for, 224–225

vaccines, 58, 74
Voynet, Dominique, *30, 31*

Waechter, Antoine, *30, 31*
"wage-earner funds," 13–14
Warren, Elizabeth, 38, 152, 224
wealth, socialization of, 9, 10, 26, 47; economic prosperity and, 124; new cycle of, 15–21; taxes on the wealthiest and, 203
wealth/income disparity, 4, 7; of municipalities (*communes*), 162; opacity in wealth distribution, 72; reduction in, 21; wealth distribution, *53*; in world regions, 87–88, *88*
welfare state, 124, 143, 155, 225
West, the, 4, 97, 101, 172, 225; military strength relative to Russia, 100; mixed economy in, 72; sharing of wealth and, 117; tensions with China-Russia bloc, 199
West Bank, 168, 196
Wilson, Woodrow, 151
women: pension reform and, 140, 143; rights of women and minorities, 77; right to vote, 121; share of labor income, 89–90
Workers' Party (Brazil), 36n20, 44
working classes/employees, 2, 17, 181; British elections and, 39; climate challenge and, 134; company shares owned by, 13,

14, 108, 109; divisions among, 105, 113, 221; ecologism and, 29; excluded from government, 219; French elections and, 164, 165; green discourse and, 34–35; progressive taxation and, 19; seats on governing bodies of companies, 188, 205; social-ecological program and, 127; urban and rural, 32; votes against EU, 109; voting rights in companies, 14, *14*, 225
World Bank, 44, 172
World Inequality Database, 4, 51, 108
World Inequality Lab, 3, 50, 172
World Inequality Report 2018, 97
World Inequality Report 2022, 87, 89, 90
World Political Cleavages and Inequality Database, 112
World War I, 5
World War II, 5, 100
WTO (World Trade Organization), 58

xenophobia, 92, 94

"yellow vests" (*gilets jaunes*) crisis, 92, 134

Zemmour, Éric, 111, 163
Zemmourism, 92
Zucman, Gabriel, 150, 152